Oil

Israel's Covert Efforts to Secure Oil Supplies

Zvi Alexander

gefen

publishing house בית הוצאה לאור

JERUSALEM ◆ NEW YORK

Copyright © Zvi Alexander
Jerusalem 2004 / 5764

Typesetting: Pardes Publishing House
Cover design: Studio Paz, Jerusalem

ISBN 965–229–317–2

1 3 5 7 9 8 6 4 2

Gefen Publishing House   Gefen Books
6 Hatzvi Street, Jerusalem 94386, Israel   600 Broadway, Lynbrook, NY 11563, USA
972–2–538–0247 • orders@gefenpublishing.com 1-800-477-5257 • orders@gefenpublishing.com

www.israelbooks.com

Printed in Israel   *Send for our free catalogue*

This book is dedicated to those who realize that brilliant ideas are only a small part of the journey to success. The other 95 percent is composed of hard work, many disappointments, and a lot of sweat and tears. I truly hope that this book will convey this message.

This book is also dedicated to my wife Rachel, our children Shaula and Kobi, and our grandchildren Eviatar, Noa, Yarden, Sharon and Ella.

Contents

Preface

For more than forty years I have been active in the oil business. In 1958, I returned to Israel, after serving four years with the Israeli Army mission in the USA. I was nominated to be the deputy managing director of Lapidoth Israel Oil Prospectors Corporation, the largest, and the only oil producing, company in Israel.

* In 1965 I became the managing director of the Israeli National Oil Company, which I headed for almost ten years.
* In 1974 I was engaged to head an oil exploration company in London, England, which I did for two years.
* From the late seventies, for more than twenty years, I continued in this field as a private entrepreneur in various parts of the world.

Oil is one of the most important commodities in the world's economy. Oil is the "engine" of the modern world. Automobiles, airplanes, power stations—all need oil in order to function. The national security of all modern nations depends on secure supplies of oil and its products.

The State of Israel, from its very inception in 1948, considered the supply of oil to be of the greatest importance. The Israeli government assigned it highest priority, second to national defense issues. Israel does not possess any alternative sources of energy; therefore, its dependence on oil and the regular and timely supply of it are paramount.

At first, oil was almost under the complete control of the Arab nations, the forsworn enemies of Israel. They made every effort to choke the supplies of oil to Israel by the use of the "Arab Boycott"—to punish any party having commercial ties with the State of Israel. One of Israel's greatest achievements was winning the war of securing regular and safe oil supplies. I am proud to have been an active participant in this war.

When I started writing this book I suddenly realized that I was fortunate in being the pioneer in several important aspects of the oil scene in Israel, details of which are described in this book. Some of these steps I succeeded in executing, despite the fierce opposition of the Israeli entrenched oil establishment whose serene lifestyle my acts endangered. Furthermore, all these successes were achieved without any financial support from the sole shareholder of the company—the Israeli government.

I will briefly list here some of those pioneering steps. In 1967 I brought to Israel, non-Jewish-owned oil companies, who together with us financed and executed the drilling of five oil exploration wells, based on projects developed by us. This effort proved that, in spite of the Arab boycott, one could find an international oil company willing to explore in Israel, if the project makes geological and economic sense. The participation of, and supervision by, such an oil company also proved that one can drill a well in one third of the time and at one half of the cost customary in Israel until then.

A year later I initiated oil exploration in various foreign countries with the participation of international oil companies that bravely joined us, in spite of the fear of the Arab boycott. Again, this world-wide effort was not supported by any Israeli governmental budget.

Another first was my approach to Wall Street to raise money for an Israeli government-owned company. One has to remember that this was more than thirty years ago, when the gap between international financial markets and Israel was larger than the distance of the earth from the moon. For an Israeli enterprise to raise money for a speculative venture that was based purely on economic considerations, without resort to any philanthropic or Zionistic considerations, was considered to be the height of folly.

My continuous fundraising efforts culminated in persuading one of the largest and most prestigious independent oil companies in the world—Signal Oil and Gas Company of California—to openly invest millions of dollars for a minority interest in our company, despite the fear of the Arab Boycott. Their participation demonstrated a recognition of our company's achievements, and it furnished us with invaluable technical assistance. Furthermore, Signal granted us an economic interest in one of the largest oil fields in the North Sea that had been discovered by them. The North Sea became one of the most important oil provinces in the world outside Arab domination.

In October 1973, after the Yom Kippur War, Israel found itself in a very difficult political situation. A large number of African countries, where most of our concessions and other interests were located, severed their relations with Israel. In addition, Signal was under a renewed attack, from other members of its North Sea consortium, to cancel our participation in the North Sea venture. They accused Signal of concealing from its partners the fact that it had granted us a five percent economic interest in their block. Signal's partners claimed that, although it was only an economic interest, the fact that it was held by an Israeli-controlled company could jeopardize their standing in the Arab world. There was a danger that Signal would not be able to defend itself and our interest might have to be rescinded.

Facing all these difficulties, the Israeli government instructed me to find a buyer for the company's assets. Despite dire predictions by government officials and several members of our Board of Directors, I succeeded in selling the assets for sixteen and a half million dollars, and additional secret rights to purchase oil. All this money, a huge sum in those days, went into the government's coffers. I do not know of any other such success in Israel's annals, where a minute investment brought such an enormous return.

The buyers of the company insisted that I continue to manage it. I was also allowed to take our senior staff to London with me. This was another first—that an Israeli executive was nominated to head an international oil company abroad.

For my numerous achievements I was officially thanked by my direct superiors, the minister of finance, Pinhas Sapir, and his deputy, Dr. Zevi Dinstein. Their thank-you letters are reproduced in this book. Nevertheless, I was viciously attacked in the Israeli press. Several journalists were goaded by members of the Israeli oil establishment who were envious of our achievements. They were afraid that they might lose the benefit of government support unless they followed the pattern established by us, i.e., to bring foreign investors into the country. These slanderers were aided and abetted by some of our company employees whom I could not take with me to London.

The journalists published various personal accusations against me, without checking their allegations. Government officials who had been so enthusiastic and so full of praise of our achievements suddenly melted away. It is quite possible that these unbridled attacks gave me the impetus to write this book.

Two people of great help to me in writing my book were Nurit Arad and her husband, Arye (Arik) Arad. Nurit was a member of the editorial staff of *Yediot Aharonot*, Israel's largest newspaper. She considered writing a book about Israel's National Oil company, describing my achievements. In 1980 she came specially to London to record my memoirs, when the events were still fresh in my mind. Her tapes have been invaluable to me. Her husband, Arik Arad, a senior editor of the newspaper *Davar*, was the only journalist who had studied the complexity of oil exploration, and understood them. Back in 1967 he had interviewed our foreign partners, who were full of praise for our collaboration, which they planned to continue in other exploration areas. Arik Arad became one of my greatest friends and supporters. It was he who had encouraged his wife to come to London to tape my story.

When I now read these memoirs I wonder how I could have pursued so many diverse activities in various parts of the world within such a relatively short span of time. Although I had very many difficulties and numerous bitter disappointments, still my efforts did finally meet with success, for which I am most grateful.

I hope that this book will open for the reader a small window into the most fascinating subject I know, the quest after "Black Gold."

Zvi Alexander, Tel Aviv, May 2004

Chapter One

The Land of Milk and Honey—
but is there Oil?

I entered the offices of Israeli National Oil Company (I.N.O.C.) at No.10 Carlebach Street in Tel Aviv, on December 31, 1965. I found fifteen empty rooms, twelve cars and two employees—a secretary and a driver. I later discovered that there were also six partially disabled drilling people on the payroll. They had been reporting for years to the central warehouse in Holon to collect their salary. The management of I.N.O.C. found it easier to continue paying them a monthly salary than to solve their disability compensation once and for all.

A month earlier, the Chairman of the Board of I.N.O.C. had suddenly been informed by the deputy minister of finance, Dr. Dinstein, that the government had decided to close I.N.O.C. and to terminate its activities. The other Board members learned about the closure from the newspapers. They read that this company, which they had devotedly served for years as a public service, had wasted many millions of pounds and was being closed for incompetence. In Israel, serving on the Board of Directors of a government-owned company did not entitle the directors to any remuneration. It was considered a public duty and a great honor.

The members of the Board of I.N.O.C. were all prominent men. They included David Touviahu, the founder of modern Beersheba and its mayor, Yitzhak Horin, chairman of Yakhin Hakal, the largest agricultural company in the country, prominent bankers and other distinguished citizens. They were all convinced that their service on the Board of I.N.O.C. contributed to the building of the State of Israel. To find out from the newspapers that their services were redundant was a very rude awakening.

Two weeks earlier, on December 15, 1965, I had left Lapidoth, the largest oil exploration company in Israel, after seven years as its assistant general manager. There was a power struggle for the control and management of Lapidoth, which I lost. In the aftermath of this struggle I had to leave.

I was nominated to a newly-created position entitled "Supervisor of government investment in the oil exploration industry." This position was created by the government *ad hoc,* so as to retain my services in the oil industry.

* * *

I started working in Lapidoth in October 1958 after returning from the USA. I had served there for four years, beginning in July 1953, as the Israeli Army Signal Corps representative in the military purchasing mission in New York. In 1953, being posted to the USA was the greatest prize that an army officer could wish for. This was a time of severe austerity in Israel. Everything was rationed. To be sent to the land of plenty was the height of success and was fought for tooth and nail. To many other candidates, the fact that I was being picked for the job added insult to injury. This job, by definition, was intended for a technical officer, which I was not. There were at least eight competitors for the position, including the signal officers of the Air Force and the Navy. All of them thought that they were more suitable for this posting than I.

* * *

The Israeli Army in those early years still basked in the glory of the War of Independence, and it was rightly proud of its achievements. But soon it became sluggish and devoid of new initiatives. The army and the whole country were very fortunate when, in early 1954, the prime minister, David Ben-Gurion, nominated General Moshe Dayan as the fourth chief of staff of the Israeli Army.

I had three opportunities to observe Dayan's thinking at close hand, and at each of them I was fascinated by his courage and originality. He always looked for new avenues to get away from rigid tradition, and usually he found them.

My first experience was while Dayan was serving as the commanding general of the southern front. He requested that Yitzhak Almog, the Chief Signal Officer, and I, the head of the manpower division of the Signal Corps, appear before him. The Southern Command was then located in a large tent near Beersheba. Without any preamble Dayan told us that he had visited the adjutant general's offices and had checked the complement of servicemen in the signal company under his command, as well as the number of soldiers in the chief signal officer's headquarters. "I found out," he said in an angry voice, "that my signal company is far below strength, while you have many more soldiers than you should have according to your table of organization."

I tried to explain to him that his Command Signal Company was composed of professionals only—i.e., wireless operators, radio technicians, cipher operators, field telephone technicians, etc.

The army had a great shortage of all these skills, while in our headquarters we employed also drivers, cooks, clerks, etc., of which there was a great abundance. "Therefore," I said, "your company is below strength, while we have a full complement of servicemen."

All my explanations were wasted. Dayan ordered: "As long as my company is 20 percent below strength, so will yours be. When you succeed in staffing my company, you will be allowed to do the same for yours." His command created a very big incentive for us to try harder and to find the necessary professional personnel for the Southern Command Signal Unit.

My second encounter was in the early summer of 1953. General Dayan was already head of army operations, the second highest ranking officer below the chief of staff. It was our custom, as heads of manpower divisions of various professional corps, to go to the General Headquarters, at the beginning of July, before the school year ended, and demand as many high school graduates as possible for each of our commands. In those early years there was a small number of high school graduates and each professional corps was fighting for the largest allocation. When we arrived at Headquarters, on this occasion in 1953, we were faced with a very unpleasant experience. We were told that, following General Dayan's orders, 50 percent of high school graduates were going to the infantry and only the remaining 50 percent would be divided among the professional corps.

Until then the infantry had been regarded as the army's "Siberia." Infantry then was primarily staffed by soldiers with little education and by new immigrants who could hardly speak Hebrew.

Dayan's directive was very much resented by the professional corps, but his brave and valiant decision in one swoop made the infantry a proud fighting entity in which the best people were serving. This decision has contributed immensely to the fighting spirit of the whole Israeli Army.

The third time I witnessed Dayan's determination was when I was already a member of the purchasing mission in New York. As the United States at that time did not allow any modern armaments to be acquired by Israel, the mission was buying World War II army surplus equipment, as well as kitchens, blankets, shoes and other such items. In early 1954, when Dayan became chief of staff, a cable from him arrived, with the following instructions: "From this day onwards, the little money that we have will be devoted to purchasing arms and signal equipment only. If we won't have shoes or blankets, the soldiers will bring them from home. No more spending on such mundane items."

In this case, as in the previous examples, Dayan's decision was clear-cut and straight to the point. It completely changed the nature and image of the army. Among other benefits of Dayan's decision was a sharp increase in the budget for the purchase of equipment for the signals corps. These purchases that I was privileged to implement, contributed considerably to the accomplishments of the signals corps in the Sinai campaign of 1956 and during subsequent years.

* * *

But let me return to the controversy surrounding my appointment to go to the US. In the end, Shimon Peres, then serving as the director general of the Ministry of Defense, had to decide who would be going. I had a long final interview with him. Peres received me very warmly, and at the end of the interview he said that, as the chief signal officer had insisted on my candidacy, he would therefore abide by this recommendation.

The night before I left for New York there was a farewell party at the chief signal officer's home. The deputy chief of the Signal Corps approached me and stated that everybody was happy. "They all still hope and pray that you will break a leg before the plane takes off. One of them may still go the States instead of you."

Several months later, during Peres' visit to our Mission in New York, he told me that he was glad he had listened to the chief signal officer's recommendation. He considered me to be the best purchasing agent the army had ever had in New York.

In the last year of my service in New York I attended classes at Columbia Graduate Business School. I had difficulty in being accepted to the Graduate School since I did not have a Bachelor's degree. Like most Israelis of my generation, I had spent all my adult years in the army, first in the British Army and then in the IDF (Israel's Defence Forces), without an opportunity to study at university.

I went to see Meir Sherman, who was the economic minister in the Israeli embassy and had an office on the same floor as our offices. I asked for his assistance. We worked out a scheme that in his letter of recommendation to Columbia he would list my military courses and claim that the education thus earned was at least as important and comprehensive as a Bachelor's degree.

Columbia accepted me on a trial basis, subject to a final decision based on my grades. Fortunately my grades were satisfactory, and I was accepted as a regular student in the Graduate School of Business.

I enjoyed my studies at Columbia University very much. However, in the summer of 1957 my replacement at the purchasing mission arrived and we went back to Israel.

My service contract with the IDF was still in force and so I went back to Signal Corps Headquarters. It was very difficult to return to a military environment after four years in the US in a totally civilian capacity. The chief signal officer, as a reward for my dedicated service, offered to send me to the "Technion" (Israeli Institute of Technology), the top engineering school in Israel, for a four-year study period. Since, in return, I would have to sign up for an additional six years in the army, I declined his most generous offer and, at age 35, I left the army in the spring of 1958.

After I arrived back in Israel I went to pay my respects to Shimon Peres. Peres at that time was the all-powerful head of the defense establishment and there was no intermediary between him and David Ben-Gurion, the Prime Minister and Minister of Defense. In many areas, Peres was actually the minister of Defense.

During my visit with him, Peres said: "Arrange your things and come back." This statement did not register with me and I did not go to see him again. Only years later did I realize that Peres had felt a responsibility towards me as one of his men. What he had meant was that I should report back, after attending to my personal matters. He would then probably have arranged a suitable position for me in the defense establishment. When I met him several years later there was a distinct coolness on his side. Perhaps he felt offended that I did not respond to his invitation.

Leaving the army without preparations for a civilian occupation was an irresponsible act. I had a wife and two small children. I did not have any specific profession, nor any savings or other means of support. I also had not secured a position in civilian life while still being on the army payroll. The army felt that career officers, after a long service, were entitled to be on paid leave until they found a suitable occupation in civilian life. I did not take advantage of this privilege.

I was still wondering what to do with myself when I met a very good friend of mine, Yitzhak Almog, who had been my schoolmate at the Herzliya Gymnasium in Tel Aviv. Almog had also been my commanding officer in the British Army, and the head of the Israeli Signal Corps who had sent me for service in the US in 1953. Almog told me of a chance meeting with a mutual acquaintance who had told him that he was looking for a young man who knew English well and had management training and experience. The position offered was that of assistant general manager (in effect, executive vice-president) of Lapidoth Israel Oil Prospectors Company. Almog told this man about my return from four years of service in the US and my studies at Columbia. They both agreed that I fitted the job description and decided that Almog would present me with this offer.

For me it was a golden opportunity to realize a long-forgotten dream. As a kid I was very much interested in chemistry and mining. I had quite an elaborate chemical laboratory in the one-room apartment my mother and I shared in Tel Aviv. I once almost blew up the building in one of my chemical experiments trying to make nitroglycerine. Had I had the financial means, I would have gone to university immediately after leaving the British Army to study either mining engineering or chemistry. Also, if the War of Independence had not claimed precedence on my time I would have commenced my studies then. The opportunity to work for an oil exploration company was a dream come true.

I had two interviews with the man who originally approached Almog. He represented the Ministry of Finance on the Board of Directors of all the three oil exploration companies. He controlled the funds allocated by the Treasury for oil exploration. Consequently, he was the most powerful director in each company. After the first two interviews he introduced me to Mordechai Chen, the managing director of Lapidoth Israel Oil Prospectors Corporation (the owners of the Heletz oil field), my future boss.

Chen had only recently been appointed to his post. He was a former member of Kibbutz Afikim in the Jordan Valley, and a dedicated member of the ruling political party, Mapai. He had previously successfully managed a large government-owned heavy transport company by the name of Tovala. He was brought to Lapidoth by the minister of trade and industry, Pinhas Sapir.

Before continuing any further, I think it might be helpful to give a brief description of the political and economic scene in Israel in the late 1950s.

For twenty-five years from the time of the establishment of the State of Israel in 1948, there were, to all intents and purposes, two prime ministers in Israel, serving at one and the same time—a political prime minister and an economic one. The historical reason for that phenomenon was twofold. One was that David Ben-Gurion, the first prime minister of Israel, was not very interested in economics. The other reason was much more important, namely the enormous responsibilities that fell on Ben-Gurion of forging a nation and creating an army. He had to accomplish this impossible task during a life and death struggle for survival against five Arab armies, which were attacking the newborn State.

The ingathering of Jewish exiles, most of them completely destitute, was another enormous task. It followed the Holocaust, the greatest tragedy of the Jewish people since the destruction of the Second Temple in 70 C.E. Absorbing and educating this multitude of people, which doubled the population of the state in its first year, and who spoke fifty different languages, was a gigantic task that defies the imagination. They came from the East and from the West, during the 1948 war and immediately thereafter. Ben-Gurion therefore left the management of the economy completely in the hands of Levi Eshkol, who was simultaneously the minister of finance in the Government and the head of the all-powerful Settlement Department of the Jewish Agency. Eshkol's trusted collaborator was Pinhas Sapir, who later replaced him as minister of finance. Theirs was a mind-boggling task: to provide housing, training and jobs for almost one million new immigrants; to create a new infrastructure for the country; to build new cities and hundreds of new agricultural settlements, roads, power stations, etc. It was almost a mission impossible. Money was very scarce while the needs were enormous. Massive Jewish aid came only later, and large-scale economic support from the US followed later still.

This unusual situation of, in a sense, two concurrent prime ministers, continued for twenty-five years, until 1974, the end of Golda Meir's tenure as prime minister. Her "economic prime minister" was Pinhas Sapir, the minister of finance.

Ben-Gurion, Eshkol, Golda Meir and Sapir were the leaders of Mapai, the Israeli Labor Party, which was in control of the Jewish population in

Palestine from the early 1930s onward, during the British Mandate period. It continued in power for another thirty years after the establishment of the State of Israel in 1948, until 1977.

Today it is difficult to comprehend the power that Mapai wielded. It controlled almost every aspect of Israel's economic and social life. Vast sums of money were entering the State of Israel every year, both as donations from world Jewry and in the form of aid from the US Government. Mapai controlled the Jewish Agency which collected and brought Jewish funds into the country on the one hand, and the various Government ministries that spent these moneys on the other. It thus held ultimate power over all important activities in the country.

The Histadrut (General Federation of Labor) was controlled by Mapai. It furnished the health services for 75 percent of the country's population. The Histadrut directly owned Tnuva, which was a monopoly distributor of most of the agricultural produce of the country, including the supply of milk and dairy products. The Histadrut also owned Hamashbir, an industrial and agricultural supply conglomerate. Hamashbir also created and owned one of the largest supermarket chains in Israel. In addition, it owned Solel Boneh, the largest building contractor; Koor, the largest industrial conglomerate; Bank Hapoalim, the second largest bank; and Hassneh, the largest insurance company. The list can go on and on.

The Histadrut also owned, jointly with the Government of Israel, the national air carrier, El Al, and the national shipping line, Zim.

Mapai's economic power was therefore immense. All appointments to both Histadrut and government positions had to be approved by the Mapai party.

When the state was established in 1948, there was only a small number of people in the country who possessed technical or managerial experience. Also, a high percentage of the able and educated young people were still in the army, fighting the War of Independence. Many of them later remained in the regular army because defense requirements took precedence.

The need to fill hundreds of important positions in the government, on the municipal scene, and in commerce and industry was very great. The supply of experienced personnel, on the other hand, was minimal. The political vetting system, which tried to assure that only people loyal to the party would fill positions of responsibility, made the shortage even greater.

This state of affairs sometimes created a situation in which a person formerly in charge of the cowshed in the Kibbutz found himself at the head of a large industrial enterprise.

The man who controlled the economic fate of the state more than anyone else was Pinhas Sapir, the minister of trade and industry, and later the finance minister. He was also one of the foremost leaders of Mapai, and probably the most influential. He was the all-powerful (though benevolent) dictator of Israel's economic life. Nothing happened in the economic sphere without his personal involvement. No appointment to any position of importance was made without Sapir's approval.

I personally had one small experience, which taught me the power of the party. When I returned from the British Army in 1946, I started working in Solel Boneh, the largest building contractor in the country, which belonged to the Histadrut. I was very much envied by my friends, because times were difficult, unemployment rampant, and finding a permanent job, especially with the strongest employer in the country, was quite an achievement.

There were five of us working in the office. It controlled the central warehouses of Solel Boneh in Givat Rambam, near Tel Aviv. Two of us were twenty-five years old, myself and another fellow. There was also an ex-stevedore working with us, a physically imposing man, named Perevoski, who belonged to a more radical party in the Labor movement, *Achdut Ha'avodah.*

One day a young man entered the office and announced: "Comrades, Comrade Shkolnik, the general secretary of the Tel Aviv Labor Council, has published a notice instructing that each employee of the Labor Union should subscribe to *Davar,* Labor's daily newspaper." There was complete silence in the room, and Shkolnik's emissary was looking for a likely target to commence his interrogation. He chose the other young man in the office and began.

"Are you subscribing to *Davar,* Comrade?" he asked.

"I live in the same flat with my older, married brother, who subscribes to *Davar,* and we really do not need two newspapers in one flat, especially as I cannot afford it," replied the young man. "What is your name?" demanded the emissary.

Perevoski, the powerful ex-stevedore, who was feared because of his bulk and fighting spirit, stood up. "Who did you say published the statement?" he asked the emissary.

"Shkolnik," the emissary replied.

"Tell Shkolnik that Perevoski said that just because Shkolnik signed the request, here in Givat Rambam's office we shall not subscribe to *Davar.* Now get the hell out of here," he thundered. The emissary ran out.

Shkolnik, who changed his name to Eshkol, eventually became Israel's finance minister. After Ben-Gurion's second resignation in 1965 he was nominated to be the prime minister, in which capacity he served until his death in 1969.

But they did not let me get away so easily. Several months later I was promoted and transferred to the head office in Tel Aviv. One day I was called to see the secretary of the union of Solel Boneh's office workers. This man, who was supposed to represent me, in actual fact represented the unions, my employers. He was sitting there together with the general secretary of the clerical workers' union. The general secretary said in a stern voice, "I hear that you do not subscribe to *Davar*?"

I explained that I had recently returned home after four years of service in the British Army, that my young wife and I lived in a tiny apartment without shutters and doors, and that we could hardly make ends meet until we paid off the various loans that we had taken to acquire this small flat.

"Listen," he said, "I have a wife and three children and my earnings are in accordance with the Histadrut wage scale, just like yours, and I subscribe to *Davar*."

I replied that I was not familiar with his finances, but that mine did not allow it. After a while he got tired of me and said: "Sign!" I made an assessment of the situation, and the various benefits I would need from our local workers' secretary, which I would probably lose, and I signed. I have not forgiven them for this coercion to this day.

The Histadrut wage scale was another peculiarity of the Israeli scene, in force both before and after the establishment of the state. It was meant to equalize all incomes in public service, and base them on family size only. In spite of this supposed equality, somehow or other, the higher the position you held, the more your standard of living improved.

Actually, there was really and truly no graft. The miracle was that for a high official, the car, the driver, the private room in a hospital, and any other amenity or service that one might think of, were supplied gratis by the institution he worked for, or via the party connection. You could then honestly claim that you were paid only according to the "Family Wage Scale"—this being the official name for it.

Sometime in 1959, we went to an evening chat show which was conducted by a famous entertainer in Israel, Dan Ben Amotz. During the evening he brought out a blackboard and said, "I have a riddle for you. Yossi's father is working for Tnuva (one of the Labor Union's enterprises). His take-home salary is 486 pounds. His monthly expenses are as follows." He started listing them on the blackboard. "Rent 220 pounds, food 170, clothing 180, education 150, piano lessons 80," and so on until he reached a figure much larger than 1000 pounds. "The riddle is," he said, "how much does Yossi's father cost Tnuva?"

One more explanation is needed before we proceed, and that is about the Kibbutz. The Kibbutz, a communal agricultural settlement, was the only place in the world where the motto "Everybody shall contribute to the community according to his ability and everybody will receive from the community according to his needs" was not only preached but truly practiced. It was the only place where this Utopia truly existed. Many idealistic youngsters of my generation passed up studying for a high school diploma and went instead to establish a new Kibbutz in the wilderness. The Kibbutzim played a very prominent and vital role in the creation of the State of Israel. They took in and shielded illegal immigrants, who succeeded in breaking through the British cordon, which did everything possible to stop Jewish immigration to Palestine. Arms and ammunition of the Hagana, the Jewish secret defense organization, were hidden in the Kibbutzim. Many Kibbutzim were established far away from population centers, close to the borders of the country, to defend the Jewish population before 1948, and later helped defend the fledgling state. Also, a high percentage of the pilots in the Air Force, the paratroopers, and members of volunteer elite units of the army were composed of Kibbutz youngsters. The contribution of the Kibbutz movement to the creation and defense of the State of Israel cannot be overestimated.

* * *

Mordechai Chen was a former member of a prominent Kibbutz, a devoted adherent of Mapai, and a successful industry executive. He was therefore the perfect choice to head Lapidoth, a company of great importance for the country's future.

Lapidoth Israel Oil Prospectors Corporation was created as a result of the merger of two oil exploration companies, Israel Oil Prospectors and Lapidoth, both of which were established in 1953. The sponsors of Israel Oil Prospectors were Solel Boneh, the largest building contractor in the country, owned by the labor movement, and Xavier Federman, a wealthy industrialist and hotel owner. The sponsors of Lapidoth were the

Mekorot water company, Ampal, a US based investment subsidiary of Bank Hapoalim (Workers Bank), and the Government of Israel. The merged company, named Lapidoth, in short, discovered the first oil well in Israel, in Heletz, in September 1955. The excitement was overwhelming and the euphoria pervasive. I remember hearing the radio commentator in New York saying, "The Land of Milk and Honey is now flowing with oil, the riches of Arabia have come to tiny Israel."

After the discovery of oil, Lapidoth immediately ordered six additional drilling rigs. All of them started to work in developing the Heletz oil field. Alas, Heletz proved to be a small oil field, with total recoverable reserves of only fourteen million barrels of oil. To put this number in the right perspective, this quantity would have been sufficient only for four months of Israel's consumption at the time. The peak daily production of Heletz, which soon declined, was 5,000 barrels. This represented only 4 percent of Israel's daily requirements.

Soon enough, the numerous drilling rigs became partly idle and the two joint managing directors of the merged Lapidoth, each one representing the shareholders of one of the former entities, stopped speaking to each other. One of the employees told me that when they needed a signature on a check, they waited for one of the managing directors to leave the office. They then quickly had the other managing director sign the check. However, each check had to be signed by both managing directors. When the other managing director returned to the office and was asked to co-sign the check, he complained that he was not the first signatory. He was then offered the ready explanation that he was absent from the office when the first signature was sought.

This "happy" state of affairs could not continue for too long, especially as all the money raised originally, had already been spent. The government then assumed most of the financial burden of additional exploration costs and thus acquired a decisive influence on the company's affairs. The two warring managing directors were asked to resign and a new, single managing director was brought in—Mordechai Chen.

As Chen's knowledge of English was not very good, and as Lapidoth's chief geologist and drilling superintendent were both American, Chen was looking for a younger man to assist him in running the company and maintaining company contacts with foreigners. This is how I—ten years younger than Chen—arrived on the scene and became the assistant managing director of Lapidoth, which in practice meant executive vice-president.

My arrival at Lapidoth did not make its department heads very happy, to say the least. All of them were older than I and had already worked for the company for quite a few years. They had aspirations to become vice-presidents and my arrival quashed these hopes. The situation was so tense that during the first several months, at the end of the working day, I would examine my car in the garage, for fear that an explosive charge might have been planted in it. I probably suffered from paranoia, but I really felt that, had I disappeared from the scene, I would have made quite a few people very happy.

For the first few years, my relations with Chen were excellent. I very much enjoyed my work at Lapidoth, and Chen and I worked very well together. My only problem was Chen's love of long-winded talk. I must have heard one particular story perhaps a hundred times, about how he was stopped by a British sergeant at the checkpost near Haifa when he was driving a truck for the Kibbutz. Also, every evening our telephone was occupied for hours by Chen's rehashing of the day's events, but, after all, it was not a high price to pay for interesting and satisfying work.

I read many books about petroleum geology and felt that I really understood the complexities of oil occurrence and its commercial accumulation. Petroleum geology is not a science. It is an art that requires the assistance of various sciences. One has to have a three-dimensional imagination to really comprehend how oil accumulation is formed in the bowels of the earth. Very few people possess such insight. Of the many geologists who worked for me over the years, only a few had this insight. The rest, many of whom were highly professional, did not possess this extra sense of combining all the information together to visualize the complete geological picture.

In my view it is not a question of special intelligence. I believe that each one of us is born with certain abilities in different fields. In some we excel and in some we fail. Those abilities range from having a good sense of orientation to being a fine bridge player, to having a knack for fixing a dripping faucet or a broken lamp, and so on. I believe that in each of those activities there is a certain inborn threshold, that an individual cannot cross, no matter how much he studies and practices. I came to this conclusion many years ago while watching bridge players. Some of them had a natural intrinsic talent for the game, while others could never rise above the average, regardless of the number of courses they attended and the years of playing experience they had.

As for myself, I know that my poor sense of direction will never improve, and that I will never be able to work on my computer without frustration. These are just two of my many shortcomings. The three dimensional (3D) vision of an oil accumulation, on the other hand, was one of the things that my make-up was attuned to. I understood it and I loved it.

During those years of my "honeymoon" with Chen I was sent on a so-called study trip to Europe to visit our various suppliers there and to study the operations of the Colombe oil field in the Paris Basin in France. Going on a "study trip" was a perk bestowed on devoted officials in Israel. In the early 1960s, travel abroad was still quite expensive, and allocation of foreign currency for it needed official approval. Therefore it had to be sponsored and justified by one's employer. The official allocation of foreign currency was pitiful. Back in 1953, a time of severe foreign currency shortage, the allocation was ten dollars. Not per day, but in total, for the whole trip. When my wife, with our two small children, travelled to join me in New York, the allocation was still only ten dollars. As that trip to New York took over thirty hours, the problem with this single ten dollar banknote was a serious one. She even had a problem changing this one banknote. Finally they gave her a glass of milk for our children for free at Zurich airport.

As mentioned, a trip abroad for study and enlightenment at company expense was a very important fringe benefit, particularly when one takes into account that all the neighboring countries were Arab, and thus off-limits to Israelis. Thus it was considered very attractive and tempting to get out of the confines of tiny Israel and to go abroad.

We were planning to go by ship to Genoa. We received almost free first class tickets, thanks to Lapidoth's extensive use of the shipping line for the importation of drilling equipment. From Genoa we planned to proceed by train to France. Several weeks before setting out, I had a visit from my father's old partner, Mr. Moshe Braverman. During the 1920s and 1930s Moshe Braverman and his brother were partners with my father and his brother, the Alexandrov* brothers. They controlled a company called Tynlas (*Las* in Polish is forest), that owned large timber

---

* According to Ben-Gurion's edict, every Israeli carrying a Diplomatic or Service passport had to have a Hebrew name. Our name had been Alexandrov, but before going to the US in 1953, we officially cut the "ov" from our name and became "Alexander."

estates in eastern Poland. These forests were originally purchased in early 1920 from a White Russian Count, Count Muraviov, whose grandfather, in the nineteenth century was a governor-general of Poland. The Governor received these forests as a land grant from the Russian Tsar. The name of Muraviov was consequently not very popular, to say the least, in independent Poland. He was *persona non grata* there. All my father's meetings with Count Muraviov took place in Danzig, which was a "free city." For the above reasons, all purchases of forest estates were transacted in US dollars, which was evident in the purchase contracts.

Moshe Braverman informed me that the British government had recently issued a law called the "Foreign Compensation Act, 1960." This new law was designated to compensate British citizens for property confiscated by the Soviets in the "Ceded Territories," that is the Baltic states and eastern Poland. These territories had been annexed by the Russians after World War II. The British Government came to an agreement with the Soviet Government to use Russian gold, deposited in the Bank of England since the time of the Tsar, for such compensation and to release the remainder of the gold to the Russians.

Our forests belonged to this category. As Braverman, my mother and I were British dependents during the British Mandate period, perhaps we could be compensated for the very substantial property we had owned in eastern Poland. I asked Braverman what documents we had to support our claim. He replied that they were all destroyed during the bombardment in Warsaw in 1939. He had flown to Poland in the summer of 1939 in a last attempt to receive permission to cut some of the forest and to sell the wood.

But all forests on Poland's border with Russia were considered tank obstacles. As war seemed imminent, felling of trees was forbidden. The inducement of a million zlotys, then a tremendous sum in Poland, which was offered to obtain a permit to cut down the trees, was finally refused by a member of the Polish general staff. The general was afraid of being accused of treason. Braverman left the documents with his brother in Warsaw, who was later killed in the German air attacks in September 1939, and the documents were destroyed with him.

I asked Braverman how he expected to process the claim without any documents. He replied that Count Muraviov might still have some. To my question as to when he had last heard from the count, he answered that it was in 1938 or 1939 and that the latter was then residing in a place called St. Jacob, near Grasse on the French Riviera. I asked how old Count Muraviov was when they last communicated, more than twenty years ago. "Quite old," Braverman replied.

As our ship would be docking in Nice as well, we decided not to disembark in Genoa and to proceed to Nice instead, to search for Count Muraviov. An elderly couple on the ship heard our story and decided to change their routing and join us in our search. They were going to act as our interpreters as we did not speak French.

We rented a small car in Nice and proceeded to Grasse, where we started looking for St. Jacob. There was no St. Jacob but there was a St. Jack, and we settled for it. I was pondering where to make inquiries and decided to look for a pharmacy. If old Count Muraviov was still alive, he must be buying a lot of medicines. No luck in the pharmacy. They never heard the name Muraviov. We were standing on this beautiful summer day, the 3rd of August 1960 (the date proved to be very significant, as will be seen later), on the streets of St. Jack wondering how to continue, when an old woman passed by. We asked her and she replied: "There, you see the three houses on the hill, this is the Muraviov property."

We drove up the hill. There was a large stone house and two wooden houses nearby. We knocked on the door of the large house and an English woman opened the door and directed us to one of the wooden houses. The door opened and a tall man, in his forties, dressed only in shorts, replied: "Yes, I am Igor Muraviov."

Igor Muraviov was a professor of chemistry at Lille University, and this was his vacation home. He had arrived there two days earlier, for his month-long vacation. He remembered my father's name well and invited us in. We spent a very pleasant afternoon together. Since both Igor and his wife spoke good English, our interpreters had nothing to do except to participate in the conversation. When we discussed the possibility of finding some old documents they said they had none. Perhaps Igor's brother in Nice might have some.

After a while Igor's wife, Tania, who was also some sort of a Russian princess, said: "Igor, when we sold the big house, was there not a large box of old papers in the garage?"

We crossed the lawn back to the big house and asked the English lady for permission to look for something in the garage. She joined us, and sure enough, at the very end of the long and dark garage there was a large wicker trunk. When Igor and I tried to lift it by the handles, the bottom fell out. We turned the case upside down and carried it thus back to

Muraviov's yard. The Englishwoman was all excited. "This should be filmed," she exclaimed several times.

The first papers, on top of the pile, were my father's letters, cancelled bills of exchange and contracts for the purchase of the various parts of the forests, throughout the 1920s. It appeared that Muraviov's original land grant was immense and that he sold only a small part of it. I took a whole stack of papers and sat all night in our hotel room in Nice sorting the documents. I remember the red carpet we had in our room, which was a real mess by morning. As I did not know either Polish or Russian—in which all the documents were written—whenever I saw papers with the names "Alexandrov" or "Tynlas," I put them aside. In the morning my wife and I drove back to the Muraviovs. We drank a toast and promised them an invitation to Israel, if and when we got the compensation. Tania Muraviov berated her husband for not emulating me and demanding compensation for their much larger holdings. Finally we both succeeded in convincing her that the French government had not enacted a law similar to the British one and that there was nobody to claim their inheritance from.

The sums involved were very large. In one of the contracts, dated 1923, a purchase price of US$50,000 was mentioned. I figured out that, in 1960, this sum, compounded at 6 percent interest, would amount to over $3,200,000. This was only one of several contracts, each amounting to tens of thousands of dollars.

After we arrived in Paris, I flew to London and made an appointment with the Compensation Commission. They were very pleasant and nice. They admitted to the fact that, as Palestinian citizens, we were entitled to the full protection of the British Crown until 1948, but alas, now that we were Israeli citizens, they did not have to protect us any more.

I talked to a law firm in London and later corresponded with them for some time. There was probably a way around this citizenship problem by forming a British company and selling it the rights to the properties. But all of this would entail substantial legal fees which I could not afford. Also the fact that British lawyers are not permitted, by law, to work on a contingency basis closed this avenue for us. In the US, an attorney could probably have been found who would have taken on the case on a contingency basis.

All that remained, therefore, from my large inheritance was a pleasant memory and a file full of documents relating to my father's estate.

* * *

In the winter of 1962 I was nominated to be the Israeli representative in a two-month-long oil exploration seminar organized by the United Nations in New York. It was perfect timing for me, as it followed my first four years in the oil business.

Those early years were full of activity, both in the field and in the office. We had four or five drilling rigs running all the time, both in the Kokhav field (a newly discovered northern extension of the Heletz oil field) and drilling wildcats around the country. We also drilled for other companies as contractors. We also almost continuously employed a seismic crew in the field.

We conducted geological planning in the office, preparing new prospects to drill. I was particularly interested in geology with which I felt a very close affinity. We had a constant stream of foreign geological consultants who came to Lapidoth for limited periods and with whom I maintained constant contact. I spent a lot of time with them during their visits in Israel and continued corresponding with them after they returned to the US. Thus these two months of theoretical studies in New York came at a most suitable time.

The UN assembled an excellent group of lecturers from both the East and the West. They came from academe as well as from industry. We had the best of both worlds. We heard lectures by the chief geologist of Soviet Russia, some of the top executives of Exxon, professors at the Colorado School of Mines—the full range.

The seminar was aimed at countries that did not have oil production, or had some, but not enough to satisfy local needs. There were about twenty-five participants, including one from Libya. Libya did not have any oil in 1962!

We were seated according to the alphabet. As the Egyptian representative was Dr. Said Amin, they placed him next to me. At first he did not know how to act towards an Israeli in such close proximity. After several days, when I poured some water into his glass (water being the only beverage we were given during the long lectures from 9 a.m. to 6 p.m.) he returned the same service to me. Slowly we became closer, and after a few weeks he confided in me that his wife had sent him a long shopping list, and that he had no idea how to go about filling it. I told him that I knew New York well, as I had spent four years there earlier, and that I would be happy to assist him. We spent the next weekend shopping and finished up in a Chinese restaurant. By the end of the seminar we were great buddies.

I also became very friendly with the Malagasy (Madagascar) representative in whose country we later explored. All in all, there were correct relations among all the participants. I enjoyed the seminar immensely and learned a great deal.

One thing that interested me at the time was geochemistry. We had a geochemical survey running in Israel at the time, organized by Dr. Leo Horwitz, of Horwitz Laboratories in Houston, Texas. In this survey you extracted, by an auger, soil samples from a depth of six feet, sealed them immediately in plastic bags and despatched them by air to Texas. In the laboratory they examined the samples for the presence and concentration of heavy hydrocarbon gases, such as butane, pentane, etc. Dr. Horwitz's theory was that, throughout the millions of years of the existence of an oil field, a minute quantity of its constituent hydrocarbon gases escapes and traces of them are trapped close to the surface. Every oil field is covered by an impervious rock layer, called "cap rock." According to Horwitz's theory, there would be a large concentration of gases at the perimeter of the oil field, where the cap rock is not very effective, in the form of a "halo," a smaller concentration above the oil field, and almost no heavy gases would be found in areas where no oil accumulation exists.

That very different concentrations of heavy gases occur in the near surface is a proven physical phenomenon. But its relevance to commercial oil accumulation is an entirely different story. I discussed this subject with the Russian scientists and other lecturers at the seminar, but I did not receive a definitive view on the subject.

After the end of the seminar I visited the research laboratory of Amoco in Tulsa, Oklahoma. I asked the head of the laboratory what he thought of geochemistry, and he replied: "It is harmful."

I could not believe my ears. "How could it be harmful?" I asked.

"Because it is meaningless," he retorted. "If you are weighing the pros and cons of whether to drill a well, the scales sometimes weigh against your drilling it. If you then consider the positive geochemical results, it may induce you to drill. This is the harmful effect. This type of geochemistry we consider meaningless. We therefore do not include it in the equation," he said.

"How does Horwitz survive?" I asked.

"You employed him, did you not?" the Amoco scientist snapped back. "How many companies like you does he need each year to make a good living?"

Horwitz Laboratories probably exist to this day and many companies probably use their services. But a direct method of finding oil, such as this geochemical survey mentioned, has not been found yet. Billions of dollars have been spent to find a method that will directly indicate the presence of oil. No luck yet.

I returned to Israel after the seminar with much more theoretical knowledge of the oil industry.

* * *

One week in particular, in 1962, remains vividly in memory. This is the week that we thought in Lapidoth that we had solved the country's energy problems.

We were drilling Kokhav number 10 in the northern extension of the Heletz oil field. We drilled down approximately one hundred feet below the producing horizon, as was our custom. Suddenly the drill bit dropped six to eight feet. We figured that we had encountered a cavern full of oil. Oil is usually found in the minute pore space between sand grains of sandstone, or in tiny cracks in limestone rock. To find oil in a kind of cavern is most unusual.

The well started to act up and there was great danger of a blow-out. A blow-out is usually a disaster. The well gushes oil or gas, quite out of control, and usually catches fire. To contain it one sometimes has to drill a relief well nearby, to reach the producing horizon. Through such an additional well it is possible to "kill" the gushing well. All this time, until the relief well reaches the intended depth, the "blow-out" well keeps burning.

We posted guards around the rig in order to prevent anybody with a cigarette or other incendiary material from approaching it and causing the escaping gas to catch fire.

One way to control the well permanently was to lower and cement a pipe (casing) in the hole. In this way the gushing horizon is sealed off. Perforations made through the pipe would then allow controlled production.

Our American drilling superintendent, Homer Dunigan, went up on the rig floor, lowered the pipe, and saved the well.

Homer Dunigan was quite a character. He was in his sixties, tireless and fearless. After his retirement in America he came to work for us. Except for his wife, Lois, who nagged him to death, he did not fear a living soul, or even a bubbling oil well ready to blow out.

He once told me his life story and how several times in his career he was promoted from driller to tool pusher (rig supervisor) and then to a drilling superintendent, supervising several rigs in different locations. During a periodic slump in the oil industry, he was reduced in rank to tool pusher or even driller. This cycle repeated itself several times during his career.

I often thought how different the situation was in Israel. The biggest crime an employer in Israel could commit was to reduce working conditions, be it the job description, or the salary, or any of the fringe benefits of an employee. Such a thing could not take place. It is called in Hebrew *Hara'at Tnaim*—worsening of conditions. It was completely and totally unacceptable.

For an employee in Israel there was only one way to go, namely up the ladder. This may have been one of the main reasons for the poor performance of the Israeli economy in those early years.

Coming back to Kokhav 10. We were elated. An American reservoir engineer who was with Lapidoth at the time estimated conservatively that the well would be capable of producing three to four thousand barrels a day, i.e., thirty or forty times the production of an average Heletz oil well. He called his boss in the States to report that, in his professional opinion, this was the minimum that Kokhav 10 would produce.

There was a good possibility that this new producing horizon—Calcerenite we called it—extended in several directions, and with luck we might have fulfilled the country's oil requirements. For a whole week we walked as if on a cloud. We considered ourselves great heroes, solving the second most important problem (the first was obviously defense) of the state, energy self-sufficiency.

When we perforated the well and started producing it, water replaced the oil after several days. The cavern was full of water with a thin film of oil on top. Our bad luck!

* * *

We had another interesting experience at the beginning of the 1960s. A Californian fundamentalist named Wesley Hancock applied for licenses in the area of the Carmel mountain range. Wesley Hancock believed that God had directed him to find oil in Israel and indicated to him the area that he should concentrate on. This was the region allocated to the tribe of Asher, one of the twelve tribes of Israel that came to the Holy Land in

1500 B.C. Wesley Hancock named his firm "The Asher Oil Company." The reasons for God's command to Wesley were three:

1. The Bible says that the foot of Asher dipped in oil (to Hancock, obviously petroleum). The fact that there was only olive oil at the time of the Bible, and this was the oil the Bible referred to, did not bother Wesley.

2. There is a story in the Bible that the prophet Elijah had a dispute with the non-believing prophets of Baal. The bone of contention was, who can make water burn. Elijah succeeded. What kind of water burns? Obviously petroleum. The dispute took place on the Mountain of Muhraka, in the region of Mount Carmel, in the tribe of Asher's domain—this was the second indication.

3. Wesley was on his way to a meeting in Denver, Colorado, which was scheduled to take place in an office located on the eighteenth floor of a tall building. Wesley pressed button No. 18, but the door of the elevator opened on the sixteenth floor instead. Wesley walked through an open door in front of him. In that office he saw a photograph of a rig drilling in Israel. This sealed the matter for Wesley. (The office belonged to one Gershon Brody who had an interest in drilling projects in Israel. This is how the picture came to be displayed in that office).

The Government of Israel was more than happy to issue a license to somebody who was going to drill at his own expense. Wesley received his "Asher" oil prospecting license in no time.

I went to see Wesley Hancock to offer him the use of one of our Lapidoth rigs. Wesley did not exactly understand what I was offering him. He thought that we wanted to be his partners in drilling the well. He said to me: "I don't think that you will agree to my articles of incorporation." I answered that his articles of incorporation were of no concern to us. Still, I was curious as to what these articles were and I asked him about them. Wesley told me his articles specified that all the profits from the oil that he would find in Israel would be dedicated to the fulfillment of Jeremiah's prophecy. This was: "I will bring my people from the north country." "My people" meant the Jews, "north country" was Russia, and Wesley Hancock, the Christian fundamentalist from California, would buy the Russian Jews from Khrushchev, the president of Russia, and bring them back to Israel.

I assured Wesley that we would not interfere with his noble intentions. All we wanted from him was his use of our rigs.

Wesley drilled two holes in the Asher area. He spent more than a million dollars of his own money, which in today's terms is probably more than ten million. Unfortunately both holes were completely dry.

While Wesley was preparing to drill the third hole in Israel, he got married. His wife said: "Either you marry me and stay in California or continue fulfilling prophecies." Wesley decided to stay in California.

* * *

I finished dictating the Wesley Hancock story on November 1, 1998. When Dijana, my helper and assistant, left at nine o'clock in the evening, I saw near my computer an old newspaper lying on top of a heap of papers. We had changed the furniture in my room and taken out a stack of old documents to be reviewed, and this newspaper was amongst them. I looked at the paper which I had not seen for fifteen years. What I saw on the front page is reproduced on the following pages.

I do not know whether to believe in extra-sensory perception. But this newspaper article that suddenly made its appearance makes one wonder about such phenomena.

The Dallas Morning News

Texas' Leading Newspaper ©The Dallas Morning News, 1983　　　　　Dallas, Texas, Sunday, January 23, 1983　　　　　B　—　···　75 C

Texas oilman stakes his faith on Israeli strike

Partner in Houston firm says gusher will fulfill biblical prophesy

By Michael Precker
Mideast Bureau of The News

ATLIT, Israel — Texas oilman Andy SoRelle is ready for the day, and it could be any day now, when his well comes in.

There will be a nearly immediate flow of non-imported oil to Israeli refineries because a coastal pipeline is conveniently close. The oil will give the struggling national economy a needed boost.

There will be extra help to guard the site — next to the magnificent ruins of a crusader's castle on a small promontory jutting into the Mediterranean Sea — from the swarms of jubilant Israelis.

There will be two celebrations, one at the site, 50 miles north of Tel Aviv, and one back home in Houston for the many fellow Bible believers who have supported the project with their cash and their prayers.

There will be press kits to explain the complicated subject to the pack of reporters who will flock to Israel to write about the new Middle Eastern gusher. And there will be more investment, more work and more miracles to make full use of God's gift to the Jewish state.

After two years and $9 million, only one element is still lacking. The steel bit chewing into rock 3 miles down hasn't turned up a drop of crude.

But SoRelle is certain it will. It says so in the Bible, he says.

Chapter 33 of Deuteronomy describes Moses, shortly before his death, blessing the 12 tribes of Israel descended from the sons of Jacob.

Of Zebulun and Issachar he says, "They shall suck of the abundance of the seas and of treasures hid in the sand."

Five verses later, in 33:24, Moses speaks of Asher:

"Let Asher be blessed with children, let him be accessible to his brethren, and let him dip his foot in oil. Thy shoes shall be iron and brass, and as thy days, so shall thy strength be."

To SoRelle, the message couldn't be clearer. The treasures to be sucked are oil. The "foot of Asher" dipped in oil is the southern tip of the land the Bible gives to the tribe of Asher — right about where the well is. The iron and brass covering the foot are two metals vital in oil rigs to eliminate sparks. And the strength of Asher will come from the oil wealth SoRelle is going to find.

"The reason I love the Bible is because it's the only thing I've found to be completely truthful and accurate, and I know the prophecies will be fulfilled," he said. "This is God's time to bless Israel, and there is no better way to bless Israel than a big oil discovery."

SoRelle, a friendly, soft-spoken man who steers clear of fire-and-brimstone testimonials, resents the portrayal he feels some have given him as a Christian eccentric wagging a divining rod or relying on visions. The 62-year-old petroleum engineer is a partner in the successful Houston firm Energy Exploration Inc., whose work in Israel is backed by first-rate technology and a skilled crew of Texans and Israelis.

But his absolute faith amid a host of problems and a legion of skeptics has been the key to one of the most unusual searches for oil since Edwin Drake started the whole business in Pennsylvania 124 years ago.

"I'm sitting on a well where very funny things are happening that I have no explanation for," said Jackie Sherman, the Israeli serving as well-site geologist, who shares SoRelle's optimism.

"Logically, we shouldn't be anywhere near here; we should have abandoned the well two years ago. We're still here because of one man's faith and stubbornness. I wish more people had it. I'm going bloody mad with all these troubles, and he's calm as ever."

SoRelle's quest dates to World War II, during which he served as a fighter pilot. In a mission over Normandy, his plane was hit, but he managed to land safely by a series of maneuvers he regards as a miracle. "God became real to me, and I became a believer," he said.

Already a success in the oil business, SoRelle first visited Israel in 1968. "I started believing I had to do something for Israel," he said. "I thought, 'What can I do?' Then I realized poor Israel needed oil."

Seven years later, after his company developed a system called "radio-metrics" to predict oil deposits by measuring isotope emissions from the ground, SoRelle wrote to the Israeli government offering his services for free.

The Dallas Morning News

The Dallas Morning News: Michael Precker

Andy SoRelle stands beside the Israeli oil rig in the "foot of Asher."

Texas oilman stakes his faith on Israeli strike

After two letters, six months and no response, SoRelle fired off a "hot letter" demanding to know why his generousity was being ignored. Soon afterward, an Israeli oil expert turned up in Houston, and SoRelle received an invitation.

In 1977, he arrived with a million-dollar machine, which failed to uncover anything promising in Israel but found a lot to like in the occupied Sinai. SoRelle applied for a license to drill. But a week later Egyptian President Anwar Sadat flew to Jerusalem, Israel agreed to relinquish the Sinai to Egypt and that was that.

"I was sure we were through," he said. "I thought, well, at least we tried."

But two years later, an old college buddy showed up in SoRelle's Houston office with a biblical map showing the foot of Asher to be farther south than SoRelle had believed, an area stretching from Haifa to Caesarea. A quick phone call to Tel Aviv confirmed that the area had never been surveyed, and SoRelle was back in business.

This time the tests were encouraging, and the license was granted. Backed by a $6 million investment, one-fourth from the Israeli government, SoRelle began drilling in February 1981. Ten months later, the crew reached 17,296 feet, far past the rig's usual depth. There was no money left and no oil. Exxon or Mobil probably would have moved on long before.

But not SoRelle. "I've learned God has a time for everything," he said. "It wasn't God's time yet."

He secured the use of Israel's biggest rig, which can drill to 26,000 feet, raised another $3 million and came back. On Dec. 28 the drilling resumed, at a cost of $23,000 a day, and last week the well passed 18,000 feet.

The project has been accompanied by a string of difficulties, and SoRelle regards the solutions as divine miracles. The radio-metrics equipment was not supposed to work on wet ground — but it did. The Israelis unexpectedly allowed foreigners to drill on a restricted naval base where their own archaeologists are not allowed to dig. More investors were recruited in the nick of time.

At one low point, SoRelle's wife, Maxine, lost a gold bracelet while swimming in the sea. A week later, when it should either have been buried on the bottom or floating toward Italy, they found the bracelet sticking out of the sand in a "V" shape — for victory, SoRelle says.

As for the well itself, "I've never had so many problems," SoRelle said. "We nearly lost it a dozen times." The problems included thicker-than-predicted rock formations, unexpected water flows, broken pipes, drills that got stuck, odd combinations of high and low pressure zones that almost caused the well to collapse.

"But all these problems were overcome by prayer," he said. "I'm not exaggerating. There are thousands of people who are praying for this well."

Most of them heard about the project through Christian news media, where SoRelle has gained increasing fame. The daily drilling reports are cabled back to SoRelle's home office and broadcast on a radio station in Houston and several others in the Southwest.

Fundamentalist Christians, many of whom are among Israel's staunchest supporters, make up the bulk of SoRelle's backers. Dallas businessman A. Ford Madison heads the second-largest investor group and also chairs SoRelle's executive committee.

If SoRelle is right, they could make a fortune. "But most of them are Bible believers who want to see Israel prosper," SoRelle said. "Believe me, there are easier ways to make money."

Skeptics of the project are numerous. Golda Meir, the late Israeli prime minister, often joked that Moses turned the wrong way after leading the Israelites out of Egypt and managed to find the only spot in the Middle East without oil.

In the past 30 years, Israel has drilled more than 300 wells, but only a handful have produced oil and none more than a paltry 100 barrels a day. The national bill for imported oil is about $2 billion a year, and most experts think the only way to reduce it is by switching to coal and nuclear energy.

Even rabbis disagree with SoRelle's biblical interpretation. In Hebrew, the original language of the Old Testament, Moses told Asher to dip his foot in *shemen*, which means olive oil. The word for petroleum is different.

Most biblical scholars have interpreted the reference to oil as symbolizing fertility and prosperity. Noga Hareuveni, an Israeli expert on the connections between nature and the Bible, explains the passage by noting that the area of Asher has the best olive trees in the Holy Land.

SoRelle has little time for doubters. He scoffs at the notion that Israel's treasures and strength will come from olive oil. And he is troubled by what he sees as the average Israeli's acceptance of an economically troubled fate.

He refuses to answer questions about what he will do if spring comes, the money runs out, the bit can drill no deeper and the well is still dry.

"I won't even think about it," he said. "We're expecting some news any day. I've never believed in anything so much in my life."

Chapter Two

The Resurrection of Israel National Oil Company

My relations with Mordechai Chen, my boss, the managing director of Lapidoth, started to change in 1963. The process began when he fired our chief geologist, Walter Randell. Randell's contributions to the company were very important, and I supported and valued him. Chen and I also had conflicting views on the philosophy of oil exploration in Israel. We strongly disagreed on the need to bring foreign capital and foreign know-how into the fledgling Israeli oil industry. Chen thought that we could and should do it ourselves. I, on the other hand, believed that we needed foreign know-how, capital and experience. Our own experience was very limited, in my view. It derived from one small oil field in Heletz and an even smaller gas field at Zohar.

Chen also resented, and rightly so, the fact that some of the Board members started to phone me directly for accurate and concise summaries of the status of our various exploration sites. What exacerbated the situation even more was the fact that the all-powerful Pinhas Sapir, minister of finance and Chen's great friend and protector, started to phone me at home for the same purpose.

More important was my feeling that all three oil exploration companies were wasting the state's meager resources on unnecessary and unjustified drilling ventures. I remember once saying that I would have much preferred it, if these misused moneys had been stolen. At least somebody would have benefited from them. Instead large sums were being wasted without benefit to anybody.

I had this odd feeling that we were drilling for drilling's sake, without sound geological justification. I realized that government money was very cheap. All one had to do was to convince the minister or director general of the Treasury. Then millions would be approved and go straight down the drain. I came to the conclusion that there must be another way.

I voiced my dissatisfaction to two other executives, who occupied positions similar to mine in the oil industry, and who were of my generation. I also expressed to some of the young Board members of Lapidoth that I did not see any chance for improvement in the present set-up, and that I was considering leaving the oil industry altogether.

Another subject which spoiled my relationship with Mordechai Chen was caused by the idea of merging Lapidoth with the Delek company. We had several meetings with Delek, at which we discussed various means of co-operation between our two companies. At the end of one of those meetings, I had a brain wave and suggested to Chen that we merge with Delek.

Delek was the second largest oil marketing company in Israel. Its already great income increased from year to year, in line with the growth of the Israeli economy. It was run on a "cost plus" basis. This method, which can only be applied in a government-supervised economy and not in a free market enterprise, calls for the price of the commodity to be determined by its costs, including overheads, etc., plus a fixed profit. In addition, there was no competition whatsoever. The market was divided at pre-determined percentages among three marketing companies: Paz, Delek and Sonol. It was a quasi license to print money—a hell of a lot of money.

Delek was one of the richest companies in Israel. Its annual profits amounted to many millions, which were subject to quite a high tax rate. Oil exploration, on the other hand, was a tax shelter whereby one could write off the cost of exploration from one's taxable income. Delek could therefore invest money in oil exploration and deduct this investment from its taxable income. A merger between Delek and Lapidoth would therefore furnish Lapidoth with all the funds needed for oil exploration, while Delek would become a full-fledged oil company, exploring, producing and marketing oil products, as customary in the western world.

Chen grasped the importance of my idea and said it was a "genius thought." He asked me to hold the idea in complete secrecy until we could plan how to execute it. I obviously acceded to his request.

Several weeks thereafter, there was a Lapidoth Board of Directors meeting, in which I usually participated. Abraham Dickenstein, the chairman of Ampal, an investment company that was a large shareholder in Lapidoth, expressed his enthusiasm for "the wonderful idea of the president of the company, Mordechai Chen, of the merger with Delek." I was sitting there with my mouth open and could not believe my ears. This was the last straw in my relations with Chen.

The idea, by the way, did not materialize due to the Finance Ministry opposition. It did not want to forgo the taxes Delek was paying. It is also possible that Chen did not know how to present the idea that this was the way the oil business operated in the western world, and Israel should function in the same way. Another possibility is that the Finance Ministry's dislike for Chen contributed to the negative result.

* * *

At the beginning of 1965 Dr. Dinstein, the deputy minister of finance and the official in charge of the country's oil industry, invited me to Jerusalem to review the whole state of affairs. After a long discussion he told me that his wish was to nominate me as the managing director of Lapidoth. I was very flattered because, after all, Lapidoth was the largest oil exploration company with a producing oil field and a substantial oil income. Some time later, Xavier Federman, a Board member, an industrialist and one of the founders of Lapidoth, phoned me and said he had heard that I was planning to leave Lapidoth. He said that I should not do so, as I would soon be nominated managing director and the Board would accept all the changes I would recommend.

In April 1965 I saw Dr. Dinstein again in Jerusalem. I asked him how long it would take to transfer control of Lapidoth to me. He said: "September." I asked: "The first of September?" "15th of September," he replied. And so the date was set.

In July, Minister Sapir phoned me at home. "I am told that you are the best oilman we have in Israel. Why don't you come and visit me?"

"I am accustomed to come only when invited," I replied.

"Come and see me, then, on Friday at 3 p.m. in the Dan Hotel."

Meetings with Sapir were an important feature of the economic scene in Israel. Sapir was a big man, physically and otherwise. Every man of stature went to see him, and he knew anything and everything relevant to Israel's economic life. His black notebook was a legend. In this notebook all the pertinent figures of Israel's economy, macro as well as micro, were listed. Meetings with him usually commenced at 5:30 a.m., on the way from his home in Kfar Saba to Jerusalem. The procedure was that you entered his car, which was driven by his chauffeur, while your car with your driver followed. Your audience usually lasted twenty to thirty minutes. The next in line was waiting at the following road junction, one third of the way to Jerusalem, and you switched places with him. The route to Jerusalem before the Six-Day-War was longer than today. Sapir succeeded, therefore, in holding three meetings on the way to his office in Jerusalem. I had two or three such car audiences. They were quite an experience.

Sapir's first question on that July 1965 afternoon meeting, at the Dan Hotel, was: "We discovered oil in Heletz ten years ago, where are we now?" This was a very statesman-like question. I am afraid that in my reply I did not rise to the occasion. I talked all the time about Lapidoth and not about the country as a whole, which was Sapir's prime concern. The meeting lasted an hour and a half. It must have been a record as far as meetings with Sapir went. He was a very able but impatient man. His usual opening at a meeting was: "Start from the end." For me to hold his attention for such a long time must have been a great achievement.

At the end of the meeting he said: "Why don't you, Chen and myself meet and discuss things together?" I replied that I did not think it was a good idea, which was not a very smart reply. He asked: "Why not?" I said: "Our views are diametrically opposed and I do not think that we should argue in front of you." "Alright," he yielded, "Come and see me again after the elections to the Histadrut (the Labor Federation)." These elections were then almost as important as the general elections.

I have many times in the past mulled over this audience with Sapir. I came to the conclusion that he must have decided at the end of the meeting, "Not Yet." After all I was an unknown quantity to him. I was not a Labor Party member, nor an ex-kibbutz member. I was an ex-army man, and in those early years there were not too many army officers occupying economic positions of importance. A long stay in America and a Columbia University education were unusual features in the resumés of government company managers. On the other hand, there was Mordechai Chen, Sapir's own protégé, a devoted party member, a kibbutz member and a trusted disciple. To replace Chen with this young Turk was risky at this point. This must have influenced Sapir's decision. Sapir was very devoted to his protégés, and this was too drastic a step for him to undertake.

Two months had passed. At the Histadrut elections Mapai, Sapir's party and the mainstream party of the Labor movement, had taken a beating. Both left and right parties of the Labor Alliance gained substantially. Chen had been very active in the election campaign and had also been in charge of transportation of voters to the polling booths.

Chen must have found out about the various contacts between the Board members, the government authorities and myself. He decided to take the "bull by its horns," especially since the country's general elections were nearing and Chen's services would be needed again. He called me to his office and, out of the blue, said: "I don't think that there is enough room for both of us in the company. Furthermore I do not plan to resign."

"I am here by specific request of your shareholders," I said.

"How can they do it? I am the managing director."

"You ask them and not me," I replied.

The situation soon became impossible. My room was next to Chen's. Mail ceased to be delivered to my office. Since I had nothing to do, I spent most of my time reading newspapers, books and professional publications.

Zevi Dinstein, the deputy finance minister, folded his sails and pointed out: "After all, Chen is the managing director." Yaacov Arnon, the director-general of the Ministry of Finance and the second most powerful man there, went to see Sapir on numerous occasions to demand that the promised change in Lapidoth be honored. I saw Arnon several times during the following, for me very long, three months. At each such meeting Dr. Arnon demanded that I should stay in my office. I pleaded: "Dr. Arnon, I cannot just sit there, soon enough they will cut off my telephone."

"You sit there. Changes must be made," he said.

Finally in early December 1965, Shimon Shapira, one of the young Board members of Lapidoth suggested to Dr. Dinstein the creation of a completely new government position, "Supervisor of Government Investment in the Oil Industry," in order to keep me in the oil industry. Thus I left Lapidoth in late December 1965.

* * *

Dr. Yaacov Arnon gave me my first assignment in my new position, namely to represent the Government of Israel in a consortium organized to drill in the offshore off Israel. This group was founded by an American named Gordon Hirshhorn. Three days after leaving Lapidoth, I found myself on my way to New York where the first meeting of the consortium was going to take place.

As I did not have any administrative facilities for buying airline tickets, getting foreign currency allocations, etc., it was all arranged for me by Lapidoth. Chen and I became friends again as if nothing had happened between us. For this attitude I admire him to this day. Once the fighting was over and the danger removed, he decided to go back to business as usual. He even gave me some friendly advice: "You have to make your mark within three months. This period will show whether you can make it on your own," he said.

I had met Gordon Hirshhorn, the leader of the Consortium for Exploring Offshore Israel, several years earlier, in the offices of Lapidoth. He came to discuss with us the exchange of seismic data, relating to our respective areas. His acreage offshore bordered our licences onshore. Geology does not recognize borders of licences. It is therefore most important to tie in data from adjoining areas, in order to create a comprehensive picture of the subsurface. All other companies operating in Israel used to exchange all their seismic information freely. But during our meeting Gordon Hirshhorn said, "but our data is confidential." "In that case ours is secret as well," I replied. He left without making an agreement on data exchange. When, several years later, he realized how important data exchange was, it was too late.

Gordon Hirshhorn was a short stocky man with bulging eyes. His father, Joseph Herman Hirshhorn, who had discovered very rich deposits of uranium in Canada, was one of that country's wealthiest entrepreneurs. He was also a famous art collector. He eventually donated his collection to the National Gallery in Washington where a special Hirshhorn wing was added to house his collection. Gordon felt that he must follow in his father's footsteps and also become a great mineral explorer. Unfortunately he did not have his father's luck or ability. His father refused to help him in his exploration efforts in Israel. Once, when an Israeli newspaperman interviewed Hirshhorn Senior and asked him why would he not assist his son, who was struggling to raise money for exploration in Israel, the father made a derogatory remark.

Gordon Hirshhorn formed a small exploration company named Petrocana and received the rights to explore offshore Israel. The company was underfinanced, always struggling to stay afloat. He succeeded in mobilizing some illustrious shareholders for his company, including Prince Bernhard of Holland, but their financial contributions were very modest. The company did not have enough funds to conduct a regular seismic survey. Instead it made a Sparker survey, which is a poor man's seismic. Such a survey shows only relatively shallow reflections and it is not very accurate, to boot.

Petrocana's Sparker survey indicated a number of structures offshore, which could contain oil or gas. These structures seemed to continue into the onshore areas licensed to Lapidoth. In order to confirm the existence of Petrocana's structures, the seismic data of Lapidoth had to be tied in. As mentioned previously, Petrocana did not have access to Lapidoth data, as Hirshhorn refused to exchange information with us.

On the basis of the structures outlined on Petrocana's seismic maps. Hirshhorn succeeded in reaching, in principle, a Farm Out* agreement with Livingstone Oil of Tulsa, Oklahoma. According to this agreement, Livingstone would henceforth fund all exploration in Petrocana's licenses at its sole risk and expense for a 50 percent interest in the licensed areas. Hirshhorn was on his way to great success. The Government of Israel had taken a 12.5 percent participation in the venture. This was the reason for my hasty trip to New York, three days after leaving Lapidoth. I had to represent the government's 12.5 percent interest at the first official meeting of Livingstone Oil and Petrocana.

When the meeting convened in New York, the Livingstone representative's first request was to see the Lapidoth seismic data, to tie in the offshore structures appearing in Petrocana's maps. Hirshhorn had been advised by Livingstone Oil, before the scheduled meeting, to obtain these data from Lapidoth. He hastily concluded a data exchange agreement with Lapidoth. Unfortunately he did not have any permanent professional staff in Petrocana who could have looked at the data and advised him on the situation. When the maps were unfolded, the offshore structures had disappeared. Lapidoth maps did not confirm the continuation of these structures onshore.

Lewis G. Weeks,** the retired chief geologist of Exxon, was Petrocana's consultant. After seeing the maps, I asked Weeks whether

---

*   "Farm out" is the sale of a specific percentage of an oil concession for a royalty, cash or a work commitment, or a combination thereof.
**   Lewis G. Weeks was a legendary geologist. After his retirement from Exxon, he discovered large reserves of oil offshore Australia in the Bass Straits. With part of the income from his royalties in Australia he created, at age seventy, Weeks Petroleum, which has become a thriving oil company operating world-wide. I first met Weeks in 1962, when he was invited to advise the Government of Israel on its oil exploration policy. I was his guide in Israel and we became great friends. He submitted a long list of recommendations. I remember two of them. One was to replace all the three managing directors of the Israeli oil companies with professional people. The second was to either accept all his recommendations, in their entirety, or forget his report. The second alternative was taken by the Israeli government. They forgot his report.

he could recommend a drilling location. He replied that he could not do so at the present time. The data would have to be studied first and further seismic surveys to be performed in the offshore. I suggested that the consortium engage the service of L.F. (Buz) Ivanhoe, an American geophysicist familiar with the geology of Israel. Ivanhoe had been, for the previous five years, a consultant to Lapidoth. His assignment was to review all the relevant seismic data and recommend a course of action. This recommendation was accepted. The meeting of the consortium was adjourned, and I returned home.

In early January 1966 Julius Livingstone, after whom Livingstone Oil was named, made Petrocana a most generous offer. Livingstone Oil offered to conduct, at its sole cost and expense, a comprehensive seismic survey offshore Israel. If a drilling location would be found as a result of this survey, the original deal with Petrocana would take effect. If Livingstone would not be satisfied as to the existence of drilling locations, the entire licensed area would revert back to Petrocana without any further rights to Livingstone. The government of Israel approved this arrangement.

The Boards of Directors of Livingstone Oil and Petrocana had their meetings on the same day in mid-January 1966. Livingstone Oil approved the offer made by Julius Livingstone. Astonishingly Petrocana's Board decided to reject it. It was clear to me that Gordon Hirshhorn was committing economic suicide. He acted like a man possessed. He told me that Livingstone Oil was out to steal his license and blacken his face. For Gordon Hirshhorn, who had nurtured this license for years, and who had put so much effort in raising the necessary funds to sustain it, even the thought that the license might not be perfect was blasphemy. I remember comparing him to a father whose only son was caught stealing. The father blames everybody, the school, the teachers, the police, etc. He is convinced that his son could do no wrong. This is how Hirshhorn perceived his oil licenses. They could do no wrong.

In February 1966 Julius Livingstone came to Israel. He was a fine elderly gentleman. He held my hand, and said: "Zvi, I am the only Jew in Livingstone Oil and I am retiring on April 1st (he sold his stock to a group headed by Baron Thyssen-Bornamisza, the famous European art collector). The other executives in the company are all nice people, but they won't go ahead with the Israeli venture if we do not sign it now. Will you, for God's sake, explain it to this impossible man?" referring to Gordon Hirshhorn.

I was coming back one night with Gordon Hirshhorn and his lawyer, Liowa Komissar, a prominent Israeli attorney, from a late meeting with the petroleum commissioner in Jerusalem. I told Gordon that I was getting sick and tired of helping him and that he was going to destroy all that he had built during the previous three or four years. "You do not understand, Zvi," he said, "they all want to cheat me." In April the Israeli government cancelled Petrocana's licenses and Gordon's saga came to an end.

* * *

When I returned to Israel from New York, after the meeting with Petrocana and Livingstone Oil, I was told that my next assignment was to conclude the winding up of Israeli National Oil Company (I.N.O.C.). "Not all investments are successful," I was told, half in jest. "This is one of the unsuccessful ones. There are offices, vehicles, stores, etc.; rent them out, sell them and complete the liquidation."

I.N.O.C. had an interesting history. It was formed in 1958 by Nachum Shamir who was a good friend of Levi Eshkol, the minister of finance who later became the prime minister of Israel. In 1958 the two existing oil companies in the country, Lapidoth and Nafta, which discovered the Zohar gas field near the Dead Sea, were already in need of massive government support.

Shamir came to Eshkol with an ingenious idea. He would form a new oil exploration company which would raise ten million dollars from the public in the form of dollar linked convertible debentures, guaranteed by the government. With the first sniff of oil, the public would rush and convert the debentures into common stock. Consequently there would be no financial burden on the government to redeem the debentures. Eshkol agreed.

Shamir named the new company The Israeli National Oil Company. The possibility that I.N.O.C. would not find any oil was not even considered. By 1965 no oil had been discovered. I.N.O.C. had spent the initial ten million dollars, plus many more millions injected subsequently by the government. The devaluation of the Israeli pound against the dollar had almost doubled I.N.O.C.'s debt. As there was no oil found, obviously the public did not convert their debentures into common stock.

By the end of 1965 Dr. Dinstein, the deputy minister of finance, became very concerned about the situation in the oil industry. Not being able to make the changes in Lapidoth's management which he considered essential, he decided to ensure at least one change, namely the closure of I.N.O.C., which was an additional unnecessary drain on government resources. How he accomplished this, I have described at the beginning of the first chapter. The task of handling the details of closing I.N.O.C. fell to me.

The former Board of Directors of I.N.O.C. resigned and a new Board, for liquidation purposes, was nominated. Dr. Arnon, the director-general of Ministry of Finance became the chairman of the Board of I.N.O.C. Dov Ben Dror, accountant-general of the government, and two other officials of the Ministry of Finance constituted the rest of the Board of Directors. At around that time, Dinstein was promoted to the post of deputy minister of defense, under Eshkol, who served both as the prime minister and minister of defense. Ben Dror replaced Dinstein as the head of the Israeli Oil Industry.

From the beginning, for thirty years, one of the highest officials of the Finance Ministry acted as the overall government supervisor of the entire oil industry. During my time, Dr. Arnon was the first supervisor, followed by Dr. Dinstein, then Ben Dror, and again Dinstein after the Six-Day-War. Only in 1978, a special Energy Ministry was established.

At the first meeting of the new Board of I.N.O.C., I was instructed to wind up the affairs of the company as soon as possible. I was permitted to employ one secretary, a driver and a geologist on a three-month-assignment for the purpose of "closing the files."

* * *

In truth I probably never intended to liquidate I.N.O.C. I was hoping that if I succeeded in keeping I.N.O.C. alive perhaps I would be able to try out some of my ideas on how to go about looking for oil in Israel. I have mentioned before that in my last few years at Lapidoth I had the feeling that we were wasting government's meager financial resources on futile exploration ventures. I fervently believed that there must be a way of bringing foreign capital, together with foreign expertise, into the country to lead the search for oil. I was sure that the search for oil could be carried out better and more cheaply. I also believed that the government should not be in business, and especially not in such a speculative business as oil exploration.

I believed that, in spite of the Arab boycott, it would be possible to find foreign investors. The Arab boycott penalized companies that were doing business in or with Israel. The Arab boycott was most effective in the oil industry, where Arab power was most pronounced.

Still, there were so many diverse oil companies in the United States, big and small, domestic or foreign-oriented, private or public, that I was convinced that among this multitude we could discover the ones which would not bow to the Arab boycott and would be willing to explore in Israel.

I thought that we needed to take a new look at the geology of Israel; that there might be some ideas based on experience gained abroad we may have overlooked; that such fresh thinking might lead us to find oil which was so desperately needed in Israel.

Had I been nominated to head Lapidoth, I could have tested all of my ideas in a functioning company with a producing oil field, drilling rigs, geologists, geophysicists and all other tools and experts to search for oil.

But beggars cannot be choosers. In I.N.O.C. I had a vehicle which could be used for the above-mentioned purposes. There were offices, geological information, cars, supplies and even valid oil licenses in good standing. The fact that the Ministry of Finance decided to liquidate I.N.O.C. did not affect I.N.O.C.'s position vis-à-vis the Ministry of Development, which was allocating exploration licenses and supervising the Oil Law and its regulations.

The wish to reactivate a government-owned defunct company is one thing, but when one's express assignment is to liquidate that company—that is another thing altogether. What complicated the situation even further was the fact that the chairman of the company's Board was the director-general of the Ministry of Finance, which had decided on the closure of the company.

My first big fight in the "resurrection" of I.N.O.C. was with Ben Dror, whom I had not known previously. He found out that I had employed the geologist on a one year contract instead of the three months permitted by the Board. I remember we had quite a heated exchange in Ben Dror's office in Jerusalem. I asked him whether he knew anybody in the Ministry of Finance who would agree to be employed for three months only. I seriously doubted that there would be even one volunteer for such a short assignment. Even one year was a very short period of employment on the Israeli scene.

Israel Eliezri was the geologist with whom I signed the one year contract. He was the senior Israeli geologist of I.N.O.C. and because of his seniority neither Lapidoth nor Nafta was willing to employ him, since each had its own senior staff. He had a wife and three children to support, and was only too glad to have at least a one year's contract of employment.

Ben Dror and I finally made up and very soon we became the best of friends.

* * *

Buz Ivanhoe, the American geologist and geophysicist, arrived in Israel at the beginning of January 1966 to review the offshore seismic information, as agreed at the Petrocana-Livingstone meeting in New York. As we had plenty of space in I.N.O.C.'s offices, I gave him a room and the necessary technical support to perform his study.

Ivanhoe finished his assignment within one month, but I asked him to stay a little longer and review I.N.O.C.'s licenses. I asked him to recommend whether, in his opinion, there were valid reasons to continue exploring in any of I.N.O.C.'s licensed areas. He asked me whether he should limit himself to the I.N.O.C.'s areas only. I said: "No, go ahead and look around as well, but stay away from the Heletz lease, which is Lapidoth's for the next thirty years."

Ivanhoe had all the country's geological and seismic information at his disposal. As there was a free exchange of information among all the oil companies operating in Israel, all this material was in I.N.O.C.'s offices just like in the offices of Lapidoth and Nafta. One day I passed his room but found it locked. I tried the door harder. I heard the key turning and he opened the door and pulled me inside. Then he locked the door again.

I asked him: "What happened?"

He replied: "I found a prospect."

"Is it good?" I asked.

"Very much so," he confirmed.

"Where is it?"

"Outside your boundary, in the Ziqlag area."

"Why the locked door?"

"If a Nafta geologist walks in, he won't see anything," he replied, "but if Murray Nadler (Lapidoth's chief geologist) walks in by chance and looks at my map, Lapidoth will apply for this area tomorrow."

The Ziqlag area was then under a preliminary permit to Nafta, which was going to expire on April 1. From their inactivity in the Ziqlag area it seemed that Nafta was going to let the permit lapse.

I asked Ivanhoe to describe the prospect on one page, with no copies, and to give it to me. He asked whether I intended to give the information to Nafta or Lapidoth. When my reply was negative, he wondered why. I told him that they had all had the raw data and information at their disposal for the last ten years but had never done anything with it. I did not want this to became another wasteful project financed entirely by taxpayers' money. If it really was a legitimate prospect, perhaps one could at last find foreign participants for it.

With Ivanhoe's summary of the Ziqlag prospect I decided first to go and see Meir Sherman who was then the managing director of the Paz Oil Company. Paz was the largest oil marketing company in the country and one of the most profitable enterprises in Israel. This was the same Meir Sherman who had helped me to enrol at Columbia University Graduate School of Business in New York.

Sherman knew Buz Ivanhoe and respected his professional judgment. Ivanhoe first came to Israel in 1960 at the recommendation of Signal Oil and Gas. Signal, a large independent oil company based in Los Angeles, California, was the secret owner of a third of Paz. (They were at the very same time producing oil in Kuwait and selling Iranian oil to Israel. Their investment in Paz was a state secret known to no more than a handful of people). At the request of Meir Sherman, Signal first sent Dr. Arthur Huey, its chief foreign geologist, to advise Paz as to the exploration prospects in Israel. Paz was a very cash rich company and Sherman wanted to know whether it would make economic sense to put some of their resources into oil exploration.

Dr. Huey came to Israel in 1958 and again in 1959. His reports to Sherman were not too optimistic. When Sherman requested Huey's advice again, in 1960, Signal became a little concerned that Huey's presence in Israel might be discovered and that this would harm it in Kuwait and in other parts of the Arab world it planned to explore. Instead, it recommended Buz Ivanhoe, an independent geologist and geophysicist not identified with any particular oil company. Ivanhoe came to Israel in 1960 and kept returning at least once a year, at the invitations of Lapidoth.

When I showed Meir Sherman Ivanhoe's summary, he asked me who owns the acreage. I explained the license situation to him. At his further urging, I suggested that I.N.O.C. submit the application in the name of Paz immediately after the current permit's expiration. He said: "But you (I.N.O.C.) do not exist."

"This is true," I replied, "but you very much exist. Maybe it will work."

I told him that I wanted once and for all to disprove the myth that because of the Arab boycott no foreign oil investor would come to Israel. If I succeed—fine and dandy; if I did not succeed, Paz would not lose anything. Sherman agreed.

I then went to see Ben Dror and showed him Ivanhoe's page. He said: "Fine, but how are you going to do it? You do not exist." I replied that I would ask Meir Sherman to front for me.

Ben Dror said: "Sherman will never agree."

"Sherman has already agreed," I surprised him.

Ben Dror said if Sherman was going to help, he, Ben Dror, would try to support it as well. "Let us see how much of a turmoil it will create, and if we can survive it."

During these first few months in I.N.O.C., I had many long sessions with Ben Dror. I was amazed at his grasp of the essential factors of oil exploration.* He very soon became my friend, confidant, and a secret supporter. His big problem was that he was in charge of government policy, which had decided to liquidate I.N.O.C. His other, equally serious, problem was that what I preached, and he came to believe in, was completely contrary to Israeli oil industry custom and procedures. My recommendations threatened directly the entrenched establishment of the oil industry and its very livelihood. It was a very dangerous game.

During these months Dr. Arnon, my chairman and the director general of the Ministry of Finance, had a sneaky feeling that something was going on behind his back. He used to admonish me at Board meetings and other occasions that he does not want me "to enter through the window." I should, according to him, enter through the front door and become the managing director of Lapidoth. "It is your task to put me there, Dr. Arnon," I always replied.

---

* During my years in the oil industry, I was amazed to discover how very few people do really understand the phenomenon of oil accumulation. Amazingly, this included a large number of oil geologists as well. They obviously knew perfectly well how to draw a geological map or write a beautiful report, but they did not have, in my view, the sixth sense of finding oil. One had to have a three dimensional perception to comprehend how all the various elements which led to an oil accumulation, such as structure or another type of trap, faulting, oil migration, timing and all other factors, come eventually together to form an oil field. Very few people possess this faculty.

My main problem in obtaining the licenses for the Ziqlag area was to ensure that nobody would suspect that it was actually I.N.O.C. which was applying together with Paz. The application was worded as follows: *Submitted by Paz using the services of I.N.O.C.* The idea was to form an impression that, as Paz was inexperienced in writing such applications, it was therefore assisted by I.N.O.C. My luck was that Dr. Arnon, who was also the chairman of the Oil Commission which allocated oil licenses, and who would never have let our application pass, was going to be at the World Bank convention in Washington, D.C., at the time the Oil Commission was going to convene. Nafta's Ziqlag permit was going to expire on April 1, 1966, and the commission was going to convene around April 10—perfect timing for us. I had a couple of friends on the commission, one of whom was Brigadier General Yitzhak Pundak, who had studied with me at Columbia, and who was part of my "conspiracy" to assist the application to go through smoothly. And it did succeed. The Ziqlag license was granted to us for a three-year period.

As soon as the license was granted, all hell broke loose. Nafta and its Board of Directors were incensed at the fact that the area was snatched from under their noses. And by whom? By a dead body. Several of my "good friends" in the industry became sanctimonious and created most of the noise. Most of the flak fell on Ben Dror. One day in May he phoned me and said that he could not take the heat anymore. He was visited at home by two of my "good friends," who warned him that if he did not stop me, they would start a newspaper campaign against him. I asked him to hold on and told him that I was coming to see him in Jerusalem straight away.

I pleaded with him that this was our only chance to find out whether we could find foreign money and stop the haemorrhage of millions on what we both felt was an exercise in futility. He finally agreed, and I respect his courage to this day.

I often wondered why I encountered such strong opposition. Perhaps my lack of diplomacy was one of my failings, and the fact that all this planning and scheming was done in complete secrecy contributed to the resulting animosity. But how else could it have been done? Had it come out into the open, it would have been squashed at the very beginning.

But the main reason for the vehement opposition was the fear that, should I succeed, the whole foundation of the oil industry in Israel might founder. God forbid, the government might even demand foreign participation as a pre-condition to allocating government monies. Such heresy must be stopped, and the sooner the better.

Immediately after obtaining the Ziqlag license, I asked Buz Ivanhoe to come back to Israel and prepare a brochure for the presentation of the project for foreign investors. When he arrived, he told me that he met in Athens airport a young lawyer from New York, named Marvin Billet, who was active in the oil business. They started chatting, and when Ivanhoe told him about his involvement in an oil exploration project in Israel, Billet said that if it was a good prospect, he could find the money for it. Ivanhoe and I immediately travelled to Athens to meet with Billet. He told us that he had an associate in Rome, a famous geologist, named John Christensen. If Chris would approve the project on geological grounds, Billet would find the money.

The following week we went to Rome, where we met with Billet and Chris. Chris approved of the prospects and wrote his own report stating that there are good possibilities of finding up to 500 million barrels of oil in Ziqlag. Obviously, both trips were approved and supported by Ben Dror.

In the following weeks we prepared in Israel a report relating to our other license in Netanya. With these two reports in hand, I was ready to go to the United States to seek prospective investors.

* * *

I had one pleasant experience in these early months of 1966, namely our negotiations with Belco Petroleum for the offshore licenses that had been taken away from Petrocana. I represented the government in these negotiations. Belco was a completely different kettle of fish. It was a rich, independent oil company, with oil and gas production in the State of Wyoming, as well as large production in South America, offshore Peru. Belco was founded in 1946 by Arthur Belfer, for whom it was named, and who still controlled the majority of the company stock, some 54 percent.

Arthur Belfer was quite an individual. He told us part of his life story during one dinner party in Herzliya. He had escaped from Poland, where he had been a major feather merchant, on one of the last ships to leave Europe in September 1939. He arrived in New York with 100 Polish zlotys, which were worthless since World War II had broken out. Luckily he had contacts in Switzerland, to whom he had previously exported his feathers, and he started trading and importing feathers from Switzerland to the USA.

One day, when America was readying itself to enter the war, a very large tender for sleeping bags was issued by the US Army. Belfer submitted a bid for the entire quantity. He had raw materials for only a fraction of it. The custom in the US was that a large number of manufacturers would participate in such a bid. Each of them bid for a quantity of sleeping bags conforming to the raw materials in his possession, and which he had accumulated in anticipation of the coming bid. When Belfer was awarded the whole contract, the other bidders were stuck with their stock of feathers without any market for them. They started calling Belfer to offer him their goods, but he refused, claiming to have enough of his own. Finally, he bought their feathers at his own price, saved Uncle Sam a lot of money, and made a nice pile for himself. What he did not mention was that, in the process, quite a few of his competitors went bankrupt.

After the war, when taxation on profits in the US was extremely high, Belfer was advised by his accountants that investment in oil exploration was a good tax shelter. His investment in oil exploration was very successful. He discovered the Big Piney gas field in Wyoming and created a thriving oil company.

John Lomax, who was senior vice-president of Belco and its head of exploration, represented Belco in the negotiations. Belco people were tough but pleasant. As part of the deal Belco requested that the three Israeli companies—Lapidoth, Nafta and I.N.O.C.—relinquish to Belco the majority interest in their licenses which were bordering the offshore areas. Thus, in the event of discovery, Belco would control both the offshore and the adjoining onshore areas. This relinquishment was accomplished after Ben Dror convinced, or rather forced, Lapidoth and Nafta to yield. He did not have to convince I.N.O.C.

When we finished negotiating, on a Friday afternoon, I took John Lomax for a tour of Old Jaffa. The poor fellow had not left his hotel room, where we had been negotiating, for the last ten days. I asked him how Belfer, a refugee who came to America in 1939 penniless, accumulated a fortune of several hundred million dollars (in today's money it would probably be comparable to several billion dollars). "He is a trader," Lomax replied. I still did not understand.

Lomax then gave me an example: "A trader is a man who will offer you fifty thousand dollars for your half a million dollar building. You will be surprised. Sometimes such an offer will be accepted."

Lomax then related to me an experience he himself had had as a young geologist working for the legendary H.L. Hunt, probably the wealthiest American of his time. Hunt was also a trader, according to John Lomax's definition. Hunt and Gulf Oil were exploring together in a certain area in

Texas. Lomax accompanied Hunt to a meeting with Gulf, where Hunt proposed that Gulf and Hunt part ways. Hunt suggested that Gulf leave all the area to Hunt Oil, in return for which he would give them a 1/64th royalty (approximately 1.5 percent of the future oil production, without Gulf making any further investments). They then left the meeting, with Hunt telling Lomax: "Let us give them time to call their bosses in Pittsburgh." When they reconvened the Gulf representatives demanded a 1/32nd royalty (double the amount that Hunt offered them). "You drive a hard bargain, boys," Hunt said, "but you have a deal." The 1/32 royalty, Lomax said, was similar to a fifty thousand dollar offer for a half million dollar house.

On the way back H.L. Hunt said to Lomax: "John, I want to congratulate you for your wonderful help in the negotiations."

"But I did not say a word, Mr. Hunt," Lomax replied.

"That is it," said Hunt. "When two parties are trading, only one man should talk on each side." On several future occasions I realized how good this advice was. Even two lawyers, representing the same party, often defeated each other's argument by trying to help.

Chapter Three

Drilling in Israel with the Americans

In July 1966 I left for the States in my first effort to raise money. Nowadays, when many Israeli enterprises are listed on American stock exchanges and when every Israeli entrepreneur of importance tries to raise funds in the States, it is hard to conceive how difficult and unfamiliar a task this was in 1966.

Furthermore, to obtain financing for a speculative investment such as oil exploration, in Israel, was almost a mission impossible. Israel had not found oil in the previous ten years, in spite of highly active exploration activity and a very substantial outlay of capital. In fact, only one oil and one gas field, both of very modest proportions, have ever been found in Israel. Comparing the small area of Israel to the vast expanses of the Arab world, with their tremendous oil deposits, in addition to their hatred of Israel with the resulting Arab boycott, made the task of raising the needed funds quite hopeless.

The Arab boycott was applied to Coca Cola, which had licensed a bottling plant in Israel. It also included Xerox which sold equipment in Israel, and a long list of other major corporations. To believe that one could find an oil company to invest in Israel was naive, to say the least.

I first went to see Marvin Billet in New York. We met one or two prospective investors but there was no interest. I left for Denver, Colorado, to see Jack Grynberg, an independent oilman who operated under the name of Grynberg and Associates (the associate was his wife). Jack Grynberg was one of the most able professional oil people I have ever met and an extremely successful money raiser (unfortunately monies for oil exploration are not investments as most of the holes drilled are dry). Jack had turned money raising for oil projects into an art. He was the first man to use mass marketing methods to raise funds for oil deals. Usually, in the oil industry, then as now, deals are made and exploration consortiums formed on the basis of personal friendship and past association, and not mass marketing.

I have not seen anybody else use the mass marketing formula in the oil exploration business either before or since. Grynberg's contention was that it is impossible to know whether any company is ready, at a given point in time, to take up a new venture. It depends on so many different factors, such as the company's current budgetary situation, whether it is overextended or looking for deals, whether it likes onshore or offshore, Africa or another continent, whether it has a new chief geologist, what was its most recent experience, and so many other considerations which one can never know in advance.

In one particular deal which I examined, Jack wrote to over one hundred companies that he considered as potential participants in the deal. He received more than a hundred negative replies. But there was one positive reply from a serious company named US Smelting and Refining. This company had recently been taken over by a group of businessmen from New Jersey who had an appetite for a foreign venture. Jack would never have thought of them had he not covered the whole spectrum.

Jack honestly and truly tried to help me. I remember one telephone conversation he had with an elderly gentleman who was reported to be the wealthiest individual in the state of Wyoming. This person had just donated three million dollars for a hospital wing in Israel to be named after him. He was an investor in several of Grynberg's ventures. I listened in on the telephone conversation. Jack was saying: "Ted, you helped me out years ago when I was just starting in business. Believe me, this drilling deal in Israel (Ziqlag) is very good. I checked it out and I like it very much. I need your name, Ted, to convince other investors. If you will commit a hundred thousand dollars, I can easily raise the remaining nine hundred thousand." The man responded: "Fine, I will let you know in a couple of days."

Three days later the man's accountant called to inform us that such an investment of a hundred thousand dollars abroad (out of a reported two hundred million that he had) might complicate the issue of his estate. Therefore they regretfully declined. I was ready to cry.

At the end of two weeks, I left Denver. I decided to go and see every Jewish oilman I had heard of. I went to Los Angeles, Wichita (Kansas), Oklahoma City and Tulsa. In most of the places I was received in a friendly manner, but I never reached the stage of showing my maps. I heard complaints about the Histadrut (the Israeli General Labor Federation), about Minister Sapir, the bureaucracy in Israel, and a hundred other things which were wrong in Israel. Many of the complaints were probably justified. But they really did not have anything to do with the geological merit of drilling in Ziqlag.

Since that time, throughout my fund-raising attempts for various diverse ventures I have tried to stay away from the Jewish community whenever the deal involved an investment in Israel. I believe that there was a problem on both sides of the ocean. On the one hand, there were then very few successful Jewish investments in Israel. Almost every wealthy Jewish businessman knew somebody who had lost money in Israel. On the other hand, I think that the Jewish people are confused when they have to judge the economic merits of a specific investment in Israel. They do not know whether they are influenced by their heart or their brain, and in order not to be confused they decide not to invest. As a result billions of dollars flow to Israel in the form of grants and donations, but much less than 5 percent of this sum comes in the form of investment. In my opinion, Israel would have benefited much more from fewer donations and more investment, even if the total sum would have been much smaller.*

During my stay in Oklahoma City on this first fund-raising trip in 1966, I was told by a certain lawyer that there was a wealthy company in town by the name of Helmerich and Payne which was ready to go "foreign." But he warned me that they were very anti-Semitic. Old man Helmrich was of German descent. He had financed anti-Jewish propaganda during World War II. I replied that I was not selling a Jewish deal, but an oil deal, and would like to meet them if they were interested.

I went to their offices at four o'clock in the afternoon and must have remained till after seven. Harrison Townes, their chief geologist, stayed with me, reviewing the geology, budgets and other related matters. He showed real interest in the Ziqlag project and promised a decision within one week. This was actually the first time during this long trip that I had seen any real interest, and my hopes rose accordingly.

Several days later Harrison Townes phoned me to say that they were

---

* This chapter was written in 1988. The situation now in the late 1990s and early 2000, is very different. There are very many successful foreign high-tech investments in Israel, both from Jewish and Gentile sources. Israel has become the second Silicon Valley.

not joining. When I asked for the reason, he said: "Had your budget per well been less than two hundred thousand dollars, you would have had Helmerich and Payne as your partners in Israel." I asked him whether he would be kind enough to explain the reasons for their refusal to participate in writing. His letter confirming the above was waiting for me in Israel.

In preparing my budget for Ziqlag, I used the lowest figure that a similar well, to a 7,500 foot depth and to the same geological horizon, had cost in Israel. This was Beersheba No. 1, which took ninety days to drill at a total cost of $284,000. And how right Harrison Townes was. A year later, together with our foreign partners, we drilled a similar well in thirty days and at half the cost.

One of the people I met on this trip asked some embarrassing questions. He said: "You are a government-owned company, right? There are other government oil exploration companies in Israel, right? So what you are trying to tell me is that the good deals you bring to us, whereas the lousy deals they drill in Israel themselves at government expense. You must be insane. The Jews are not so dumb."

* * *

I arrived in New York on a Friday, five weeks after coming to the States. I was completely dejected. My great hopes of proving that money could be found for exploration in Israel, and that the situation in Israel could be changed, proved to be a pipedream. I would have to admit failure. I felt sorry for Ben Dror, he would be ridiculed for supporting my dreams. I called Marvin Billet from the airport to say good-bye. Marvin told me that the American Petroleum Institute (API) convention was to get underway in New York the following Monday and that it was worth a last try to find prospective participants for the Ziqlag venture there. I decided to stay, and on Monday morning Billet and I arrived at the New York Hilton, where the convention would take place.

The first man we met in the lobby was Kenny Bordo, who was the owner of oil field equipment depots all over the Middle East. "Kenny," Marvin said, "can't we help this poor guy? He really has good drilling prospects. Chris (our geologist friend from Rome) thinks highly of them. Who, in this convention, can we talk to that is not afraid of the Arabs?" Kenny consulted with the president of Reed Drilling Bits, who was standing nearby, and then advised us to talk to three people: Ray

Christian from Oklahoma City, who had gone to Libya and was disenchanted with the Arabs; John Mecom from Houston, whose drilling rigs had been confiscated by King Hussein of Jordan, and the new president of Champlin Oil, who was looking for new ventures.

We went to the registration desk and found out that Ray Christian was staying in room 654. When we called him, he invited us to come up. We found a good-looking man, some five years younger than myself, in a dressing gown. He invited us to sit down and tell our story. Ray looked at the maps, reviewed the budget and appeared very interested. He said that his vice-president, a petroleum engineer named Bill Van Meter, was then en route to Louisiana. If he would approve the technical side of the project, then we had ourselves a partner. We met Ray again that evening in one of the hospitality suites of the convention, and he and I had a long and pleasant conversation.

The next day Ray and Marvin came to my hotel. Ray said: "I understand that you are already forty days on the road and that you are very eager to go home. I can assure you that you will hear from me within a week. You can go home now." With this assurance I left that evening for Tel Aviv.

* * *

Ray Christian's letter arrived within a couple of days of my arrival in Israel. He wrote that, subject to their review of the situation on the ground, Mayflower, his company, would become our partner in the Ziqlag project. Bill Van Meter, the vice-president of the Mayflower Company, accompanied by Marvin Billet, arrived in Israel in November 1966.

Van Meter, who is today one of the top executives of the oil industry in the US, received a shock during this visit. He expected to find a desert full of camels. Instead he found a civilized country, with roads, telephones, people speaking English, and overall conditions not so different from his native state of Oklahoma. Bill told me, much later, that when Ray Christian sent him on this mission he considered Ray to be out of his mind. But Ray was paying for the tickets. Now he had found the surprise of his life. Van Meter spent several days on the derrick floor watching the drilling operations in Israel. He was impressed with the abundance of equipment and the quality and dedication of the drilling personnel. He repeated several times that all the Israeli personnel needed was technical direction and good supervision. How right he was we found out nine months later.

In March 1967 I was invited to come to Oklahoma City to sign the agreement with the Mayflower Company. Meir Sherman, who headed Paz, which owned 50 percent of the Ziqlag license, empowered me to negotiate and sign on his behalf as well. Adi Ephrat, our legal advisor, and I spent ten days in Oklahoma City. This was one of the most interesting and relaxed periods of my life. The pleasure of civilized negotiations, based on logic and mutual goodwill, was a revelation to me. This was a very different way of negotiating and doing business from that which I was used to.

In the meantime, Ray Christian was preparing a presentation brochure for his prospective investors in this deal. It was, and still is, customary in the States, to organize a group of investors to participate in and finance a drilling venture. Because of the diversity of such ventures and the relative ease of organizing them, without any government interference, the United States had more operating drilling rigs than the rest of the world combined.

In Ray's presentation, the cost of drilling the Ziqlag well was budgeted at two hundred and forty thousands dollars, with the work to be completed in forty-five days. When I saw this, I said: "Ray, you have never been to Israel. You do not know the conditions there. Why are you doing it? Why are you proposing such an unrealistic budget? Why are you promising to finish drilling the well in half the usual time?"

When he got tired of my nagging he said: "You know what, why don't you go Turn Key?" A "Turn Key" contract means that a participant pays only the budgeted figure. If it costs more, the drilling contractor pays the overage; if it costs less, the drilling contractor pockets the difference. When I came back to Israel and told Meir Sherman the story, he decided against going Turn Key. He said that if Christian was so sure of his estimates, he probably knew what he was doing. We went heads up with Ray Christian and saved another 40 percent of Ray's budgeted figure.

The agreement called for Mayflower and I.N.O.C. to be joint operators (managers of the venture). I.N.O.C. was to be in charge of geology and administration, while Mayflower would supervise the drilling and other technical operations. Mayflower took half of the deal, Paz 25 percent and I.N.O.C. 25 percent. As we managed to include I.N.O.C.'s past expenditures, as our contribution to the project, and as we also had casing and other equipment, necessary for drilling the well, in the warehouse, we did not have to invest any cash in the project. This was the "first time in 2,000 years" (an expression used in Israel to indicate a pioneering venture, i.e., the first occurrence since the Jews' dispersal by the Romans in 70 C.E.) that a foreign oil company was going to explore in Israel in an Israeli-developed prospect.

When the contract was almost completed, Ray said that we have to go to Tulsa, one hundred miles north-west of Oklahoma City, to see a lawyer by the name of Jerry Shuman. Shuman was supposed to provide a group of investors for Mayflower. Ray owned a small plane, and Bill Van Meter, a former pilot in the Marine Corps, flew the aircraft. We arrived at Jerry Shuman's home and started to go over the contract. When we came to the clause stating that the money would be transferred to Israel periodically and be disbursed by us, Shuman interrupted: "No way. You are not going to be the sole signatory on my investors' money. No way!" he repeated. "Each check will be co-signed in Oklahoma."

I was getting really upset, as it was almost impossible to manage an operation in this manner. Suddenly I felt someone tapping on my shoulder. I turned around, and there was Ray asking me to go to the next room. We went out, and Ray said: "Look, I need his money (in the end Shuman did not succeed in bringing any funds whatsoever), and we have to play along with him. What I will do is pre-sign checks and give them to you; then you won't have any delays in operations."

I believe that, to this day, thirty-five years later, there are "Ziqlag Joint Ventures" checkbooks, pre-signed by Ray Christian, in I.N.O.C.'s vault.

It was getting dark when we finished the meeting, and I asked Van Meter whether he could make it in the plane. He replied that as long as he could see his way it would be all right. By the time we arrived at the airport, the weather had turned really nasty. We took off. There was rain and thunder all around us, and I was wondering what the hell I was doing up there, among the elements, in the sky. Half way to Oklahoma City, I heard Van Meter calling on the radio: "Weevoca, Weevoca (a small town between Tulsa and Oklahoma City) this is 4X4, we need assistance" (or words to that effect). But there was no reply. When we arrived in Oklahoma City forty minutes later, I wanted to kiss the ground. The next day, when I opened the newspaper, I discovered that there had been a tornado in Weevoca exactly at the time we were passing over it.

A few days later we had to go to Tulsa again. Ray was getting ready to take the plane, but I said: "Ray, you may take the plane, you can take a boat, but the only way I am going is by car."

"You insist?" he asked.

"I certainly do," I replied.

It took exactly one hour and twenty-five minutes by car, door to door.

I heard from Jerry Shuman again about three years later. On December 31, 1969, I was lying naked on my bed in the Intercontinental Hotel in Accra, Ghana, with the air-conditioning going full blast. That morning we had signed a concession agreement in Ghana and I was feeling very happy. The phone suddenly rang and I was wondering who in the world that might be. "Zvi, this is Jerry Shuman," I heard from the other end. It took me a while to remember who it was; then I said: "Jerry, I am in Accra, Ghana." "I know where you are, I called your office, I called your home, I found out that there are only two decent hotels in Accra, and I found you."

"What can I do for you, Jerry?"

"Zvi, we have a wonderful group of investors here. We would like to explore in Sinai, and we would very much appreciate your help."

I realized then that, if somebody really wants to get in touch with you, he will eventually find you. Even if it is New Year's Eve and you are in the heart of Black Africa.

The Ziqlag contract with Mayflower was signed on April 3, 1967. It stipulated that Mayflower's investment, approximately half a million dollars, had to be deposited, in escrow, in the bank by May 31, 1967. Around May 18 or 20, Ray Christian phoned and said: "Tell me what is happening. The television and radio are full of news about Israel under the threat of Nasser, the Egyptian president. What is it?"

"Forget it," I replied, "it is just propaganda."

A few days later he called again and said: "The investors are phoning. On their television screens they see the Egyptians marching and shouting, that they will destroy Israel. And you say it is nothing?"

I said, "You know what, come over and see for yourself."

"I am coming," he said. He was supposed to have arrived on Sunday, June 4. War broke out on Monday. On Saturday, June 3, after Moshe Dayan became the minister of defense, everybody realized that war was imminent. I picked up the phone and told Ray: "You know what, maybe it is for real, why don't you delay your trip for a few days."

The results of the Israeli victory in the June 1967 Six-Day-War are well known. After the military triumph over the Arabs, the war against I.N.O.C. started in earnest. The opening salvo was a series of five articles in *Ha'aretz*, the most prestigious newspaper in the country. The gist of all these articles was an attack against the silly and harmful attempt to bring foreign investors into the country and hand over to them the vast wealth of the country's mineral resources. To let the foreigners do what we are so expert in doing ourselves would be the height of folly.

One famous statement in these articles was made by Mordechai Chen, the managing director of Lapidoth and my former boss. He said: "Our boys are performing wonders with the Mirage jets. Can anyone therefore claim, in honesty, that such a simple task as drilling for oil Jews cannot carry out successfully? We do not need the foreigners. Let them stay at home."

Another article stated: "If Zvi Alexander is such an expert in raising money, let him sit in America and do it from there. We want him to stop stirring up confusion here. We have to stop this nonsense once and for all. Foreign money, investors, foreign experts, constant foreign travel. Enough of it."

The post-war euphoria and the heightened nationalistic feelings presented a heaven-sent opportunity for the oil establishment to remove the danger of a change in government policy. Such a change might have taken place following my success in getting Mayflower to explore in Israel.

The statement by Mordechai Chen that our boys, who so wonderfully fly the Mirage planes, would surely know how to drill for oil, pleased the public because it played up to their nationalistic pride, but it was completely misleading. We had chosen the best and the brightest to train as pilots. They knew that without their skills and dedication the country might not survive. They had fought in four wars. They were more experienced, by far, than any of their counterparts, anywhere in the world. None of these conditions or attributes existed in the drilling community of Israel. All this community had seen in the previous ten years was dry holes. They were always afraid that the well they were drilling would be the last one. They feared that funding for the futile exploration effort may suddenly dry up.

I tried to defend my thesis in several newspaper interviews, one in which I recounted my meeting with Mason Hill. On one of my trips to the United States in early 1967 Ben Dror asked me to find a world-class expert to advise him prior to approving the exploration budgets of Lapidoth and Nafta. After several inquiries I was advised to see Mason Hill, who was at the time the chief geologist of Atlantic Richfield Oil Company. He was the man credited with the discovery of the Prudhoe Bay oil field in Alaska, the largest discovery in America in the last fifty years.

Mason Hill was a trim man in his early sixties. As he was due to retire soon, I was hoping to persuade him to accept an assignment in Israel. Regretfully, he refused. During our long conversation he explained that one of the most difficult tasks in oil exploration is the evaluation of a prospect. But there was one sure way to succeed, and it was to find a partner for the project. He said: "When you develop a prospect, you fall in love with it and tend to sweep under the carpet any adverse information that does not fit your rosy picture. Your partner is not in love with the project, and his judgment is completely objective. When he approves it, you may be doubly assured that the project is legitimate. The fact that you always see a group of companies exploring together is not because they lack the resources, although sharing the risk is also an important consideration. The main reason is to have several different groups of experts continuously evaluating the on-going results of the exploration effort in order to eventually discover the elusive oil field."

In another newspaper interview I tried to convey my feelings about the government's role in business, especially in such a specialized and speculative business as oil exploration. I related to them an amusing true story that happened to Mordechai Chen on his trip to India. Chen went to India in the late 1950s to try and find work for our idle drilling rigs. Already then we had superfluous equipment. He visited the Indian Oil and Gas Commission, which was located in Derha Dun, in the north of India. The Indians were then drilling a thousand kilometers away, near Bombay. A telegram arrived in Derha Dun that they had lost the drilling mud. When there is a very porous section in the well, the drilling mud which fills up the bore hole escapes into the porous formation. This is called "Loss of Circulation." A telegram came back from Derha Dun: "Inform the police immediately." This is a good illustration of why a government should not be involved actively in managing oil explorations.

I told the newspaperman that, in my view, the government's role should be limited to making laws governing oil explorations and supervising their execution. Government should not be an investor, and certainly not the sole provider of funds. It does not have the means to supervise and handle business investments.

Government should instead support the industry by providing specialized subsidies in the form of roads, water supply, security and other services which sometimes make the oil search in remote locations too costly and uneconomical. In another newspaper interview I expounded on the role of the foreign investor and replied to some of the statements made in the five slandering articles in *Ha'aretz*. I said that the foreign oil investor not only saves government resources which are so badly needed for defense and social services, but also executes the task at a fraction of the cost. Another important benefit was the expertise that the foreign investor provided. He had seen hundreds of oil fields, in many diverse areas, and in different geological settings. In contrast, Israeli oil geologists, however able, in all their years of working in Israel have seen only one oil and one gas field, both of very modest dimensions.

I also explained that if oil would eventually be found, the benefits to the government would be multifaceted. It would obtain a secure supply of energy, it would gain employment for the population, receive royalties, taxes, and many other advantages. There was therefore no need for government to invest as well.

Finally, I expressed the opinion that if foreign investors are not found it might possibly mean that the project was not worth drilling at all. Raising investment money represented a sanity check on what might otherwise be totally unfettered government investment.

But I was only one voice against a large, well-entrenched industry, whose cosy livelihood I was threatening.

* * *

As a result of Moshe Dayan's appointment as minister of defense prior to the outbreak of the Six-Day-War, Dr. Dinstein returned from the Ministry of Defense to the Ministry of Finance. He was nominated deputy minister of finance, under Pinhas Sapir and returned to his former position of being responsible for the entire oil industry in Israel.

Several days after the end of the war, we rode in Dinstein's car from Jerusalem to Tel Aviv. We drove on the newly opened route, through Latrun, which had been occupied by the Jordanians since 1948. This is the only time that I remember Dinstein himself doing the driving, and not having his driver with us. He probably did not want anyone to listen in on our conversation. Until Latrun, which was approximately half way to Tel Aviv, I told him about Ziqlag, about Ray Christian and Mayflower, about Paz's participation and our plans to start drilling. He listened for half an hour without interrupting me.

When I finished my story, as we were approaching Latrun, he said: "It is a fantastic story, Zvi, you have performed a miracle. But you will never be forgiven. You made Sapir and myself look like fools. We closed the company and you opened it. Why in the name of God did you do it in I.N.O.C."

"It is a very good question," I replied. "Where did you expect me to do it, from 43 Yehonatan Street in Zahala? (a suburb of Tel Aviv where we lived and which was originally built for army officers)."

After a long discussion he said: "Start drilling as soon as possible. When the drill bit turns in the hole it is very difficult to stop it."

It was already difficult to stop me at that juncture. To turn back foreign investors, non-Jewish oil people from Oklahoma, who were willing to spend their own money on drilling for oil in Israel was a bit too much, especially in Israel, which was so thirsty for foreign investment and so desperate to discover oil.

It was twice as difficult for Dr. Dinstein. He was the founder of the "Third Arm Movement," which was created to encourage foreign investments in Israel. He had representatives, solely for this function, in every important Israeli embassy around the world. He knew very well how difficult it is to bring foreign investors to Israel. For him not to support such an effort was unacceptable.

I would like to include a short description of Dr. Zevi Dinstein who was my ultimate boss for the following seven years. He was a good-looking tall young man. He received his doctorate in law in Switzerland in 1948, where he met Pinhas Sapir, the future minister of finance, and he became his disciple and follower for the next twenty-five years. Sapir was in 1948 in charge of finances for the Jewish effort to buy arms in Europe and smuggle them to Palestine for the impending War of Independence.

Dinstein imitated many of Sapir's habits. He tried to cram as many meetings into a day as possible. He would often come to Jerusalem for meetings in the morning, return to Tel Aviv for other appointments, and go back to Jerusalem in the evening for a meeting of the Knesset (the Israeli parliament) of which he was a member. He always pushed his driver, Danieli, to the limit. Faster, faster, he would urge. He adopted Sapir's custom of meetings in the car to utilize every minute of free time. He was used to making instantaneous decisions, to many of which more time should have been devoted, so as to weigh the pros and cons. He was very different in this respect from Dov Ben Dror, my former boss, who used to take the time to study a subject thoroughly and grasp all its ramifications and only then form a final opinion.

Dinstein would always return a phone call the same day, no matter how late. Apparently he could not go to sleep with an unreturned phone call "weighing on his conscience." He was a loner who had several assistants over the years, but they all left after a short time, because he was not able to delegate authority. He was a difficult and fascinating man to work with.

* * *

There was a ban on flights of American citizens to the Middle East during and immediately after the Six-Day-War. The Mayflower delegation arrived on the first available flight after the ban was lifted. They were ecstatic. They could now travel all through the West Bank, up to the shores of the Jordan River, instead of in a country ten miles wide. We took them to the Golan Heights where they climbed on the captured and abandoned Syrian tanks. Barth P. Walker, the chairman of Mayflower, who was also their legal adviser and second largest stockholder, after Ray Christian, enjoyed himself immensely. He ran up and down the hills of the Golan. For him, his first visit to the Holy Land was a revelation and a spiritual uplift.

In September 1967 the drilling of Ziqlag "1" started. Mayflower sent three people to supervise the drilling operation—two drilling super-intendents and one mud engineer. All they brought with them was one black book in which the whole procedure for drilling the well was spelled out in detail. They were pleasant and well-mannered. Whenever they asked for some task to be performed, and were told by the Israeli drillers that they were used to doing it differently, they said: "Please, do it our way, just this time."

Miraculously, there was no "fish" and the resulting tedious and costly fishing operation, neither in Ziqlag nor in the three wells we drilled subsequently with Mayflower in Israel. A "fish" occurs when the drilling tools are caught in the hole and you have to fish them out. Such an operation may take a couple of days or even a couple of months. Sometimes the entire hole may have to be abandoned. During all the time of fishing, the full daily cost of drilling the well continues, plus the additional cost of special fishing tools, etc. We Israelis are world experts in fishing. Then our natural ability to improvise and innovate comes to its full expression. The big trick in drilling is to avoid the fish, and in this Mayflower representatives excelled. They knew exactly when to change the drill bit, not too early and not too late, but just in time.

Changing the drill bit is a very expensive task. You have to come out of the hole with the full drilling string sometimes six thousand feet or more long, and unscrew each piece of pipe, in order to exchange the drill bit, which is obviously at the bottom. All this time, which may take a whole day or more, the full cost continues to run. On the other hand, if you let the bit drill too long in the hole, it breaks up, with a resulting fish. It takes a lot of experience and a sixth sense to know when to change it.

Another very important function was fulfilled by Mayflower's mud engineer. The drilling mud performs many essential functions in drilling a well. Mud is heavier than water and therefore keeps the fluids in porous formations, whether water, oil or gas, from gushing out of the hole with a resulting blow-out, fire etc. The mud also creates a kind of plaster on the walls of the hole to prevent cave-ins. It lubricates and cools the drill bit which heats up while constantly breaking the formation rock. Finally the mud, by being a viscous colloidal liquid, brings up to the surface the broken pieces of stone, which are then examined for formation identification and for signs of oil or gas. It is therefore extremely important to have the correct mud weight and composition appropriate to the depth, pressure and rock formation being drilled through. The mud is always in circulation, pushed into the hole by tremendous pumps on the derrick floor. Mayflower sent a graduate engineer for mud planning and supervision. We, in Israel, had a half-baked technician for this task.

We did not find any oil in Ziqlag "1." We had some non-commercial shows of gas. But we finished the complete operation in almost one third of the customry time in Israel, in thirty days and six hours and at 50 percent of the cost. The total cost was $146,000.

After the drilling at Ziqlag "1," I.N.O.C. became a fact. Perhaps an unpleasant fact, but it was here to stay, at least until it slipped up and a new attack could be launched against it. But as far as the government budget was concerned I.N.O.C. did not exist. I.N.O.C. did not receive any government money for the whole of the next seven years, throughout my services as head of I.N.O.C., even when we had world-wide operations. I broke the rules of the establishment's game and I had to pay for it.

Minister Pinchas Sapir's attitude was much more complicated. He had ambivalent feelings toward I.N.O.C. On one hand, he was annoyed that I supposedly made a fool of him, by reactivating I.N.O.C., which he had declared closed. This feeling of irritation was continuously reinforced by every executive of the oil industry that came to see him. Each one added a little poison. On the other hand, he was a true Zionist believer and a smart man to boot. How could he close a company that brought in foreign investors, drilled for oil, and did not cost him a penny? Therefore a kind of truce was declared. Let I.N.O.C. exist, but let them eat grass. No budget presence and no money.

In the meantime, we hired some additional professional personnel, which we needed to fulfill our side of the operating agreement, and to develop some new ideas. The way we survived was by charging Ziqlag and subsequent ventures with direct costs, overhead and operators fees. Another source of income were the fees we charged to the venture for past seismic work done by I.N.O.C., which we claimed was necessary for the present endeavor. This increased our contributions to the venture and we did not have to pay any cash calls. All of it was not much money, but it was enough to "pay the rent."

Following the findings in Ziqlag "1," the participants in the Ziqlag venture decided on additional seismic work in the licensed area. The seismic survey indicated a large new structure which appeared to be much higher structurally then Ziqlag "1." Oil, being lighter than water, usually migrates to a higher place; therefore a structurally higher location should be generally more prospective. The drilling of Ziqlag "2," in the new location, was agreed upon.

After we finished drilling Ziqlag "1," Ray Christian visited Israel for the first time. Ben Dror, who was the chairman of I.N.O.C., took us to see the Minister Sapir. When we stood on his office veranda and viewed the panorama of Jerusalem, he asked Ben Dror in Hebrew: "What did you give him?" (What Sapir meant was what kind of special incentive often promised to foreign investors to encourage them to spend money in Israel.)

"Nothing," replied Ben Dror.

"Come on, tell me the truth," insisted Sapir. It was inconceivable to Sapir that a non-Jewish oilman from Oklahoma would come and spend his money on the merit of it without ulterior motives, without a special grant or subsidy.

Moshe Ettinger, the petroleum commissioner (he later went to work with Jack Grynberg, in Denver, Colorado, and remained with Grynberg for many years) came to see me before Christian's departure from Israel. Ettinger wondered why I did not arrange an interview for Christian with the economic editors of the daily newspapers. Here was a genuine, honest to God, oil investor from Oklahoma and I did not have the sense to build on his visit to improve my public relations!

As Christian was leaving the next morning, we hastily tried to arrange a press conference at the airport, before his departure. It was very difficult to wake up Ray Christian that morning. Israel Eliezri, our chief geologist, must have driven at a hundred miles an hour to Ben Gurion airport, so that Ray Christian would not miss his plane. Luckily the plane was late and we were able to hold the press conference. Several newspapermen were there, including Arye Arad from *Davar*, who was also a representative of the news agency Itim. Arad's half-page-long interview therefore appeared in almost all the newspapers. In it, Ray Christian said that he had had a wonderful experience in Israel. He had met excellent and dedicated people, and his staff had immensely enjoyed working with the Israeli drilling personnel. He said that drilling a dry hole is par for the course. Only one attempt in ten or fifteen around the world finds oil, and only one in fifty discovers an oil field. If every hole found oil there would not be an oil industry in existence. He said that he was going to drill Ziqlag "2" with I.N.O.C. and that he was seriously considering joining I.N.O.C. in Netanya, our other license. He added that, compared to other experiences he had had around the world, Israel was a paradise. Every man present sat with his mouth open.

Shortly after Ray Christian's visit to Israel, Dr. Dinstein convened a one-day seminar in Jerusalem to discuss the state of oil exploration in the country. The participants were all the senior executives and Board members of the three oil exploration companies. That same morning an interview with Dinstein was published in *Davar*. Dinstein told Arye Arad, the journalist who had interviewed Ray Christian at the airport, that the only correct way to go about oil exploration in Israel was that of I.N.O.C. Israel must find foreign investors with oil expertise. "This is the only way that some sanity can be introduced in Israel's exploration efforts," he said. He also remarked that the big difference between a Jewish non-oil investor and a gentile oil investor was that the gentile blamed neither the state, nor socialism or Zionism when the result was a dry hole.

When the seminar in Jerusalem opened that morning, one could feel the electricity in the air. One of the participants asked a point of order question. "If the views published in *Davar* and attributed to Dr. Dinstein are really his, why is there a need to continue this meeting." Dinstein replied that the meeting would consider all the various relevant issues and asked the questioner to sit down (and shut up).

There were two strong attacks at the meeting directed against I.N.O.C. and me. Yitzhak Eilam, the former chairman of Koor (the largest industrial conglomerate in Israel, owned by the General Federation of Labor), who was also the chairman of Lapidoth, was the first attacker. The other was Israel Dickenstein. He was a successful industrialist and a great friend of Pinhas Sapir, whom he went to see every Saturday morning. Dickenstein was very active on the Board of Nafta, to which he had been co-opted by Sapir, as a trusted member representing the public. His foremost interest was the wellbeing of the State of Israel. Dickenstein stated in his attack that I.N.O.C. had terminated the drilling of Ziqlag "1" prior to reaching the producing horizon. This "crime" had been committed just to please Zvi Alexander, who wanted to prove that he could drill faster and cheaper. For a man of Dickenstein's caliber to express such nonsense, to accuse foreign investors of having wasted their money and stopped drilling to satisfy some foolish whim of Zvi Alexander's, was a height of folly. It frightened me. It showed the depth of hatred that must have been instilled in this man by his colleagues on the Board of Directors of Nafta.

Some time later, one of my friends met Israel Dickenstein at a private gathering. The conversation turned to oil exploration and my name was mentioned. Dickenstein said that Zvi Alexander was the black sheep of the oil industry and its destroyer. My friend suggested that I meet with Dickenstein. He added that I could not afford to have Dickenstein, who had Sapir's ear, as my foresworn enemy.

I phoned up Dickenstein and suggested a meeting. He invited me to come and see him at the aluminum factory "El Bar" that he managed and partly owned. The plant was in Kfar Saba, the town where Pinhas Sapir lived. Dickenstein added, on the phone, that I should know he would continue fighting me as long as I would continue to be the black sheep of the "Oil Family." What a 'family'! They were ready to destroy each other individually, and collectively finish off I.N.O.C.

Dickenstein was a good-looking man in his fifties, well dressed and softly spoken. He received me graciously, and I suggested that we concentrate on the issues and not speak about personalities. He agreed.

I told him about my visit with Mason Hill, the geologist of Atlantic Richfield, and how I learned about the importance of several different geological teams judging and approving a drilling prospect. I related our experience with Mayflower and the resulting cost savings. Then I described how Ray Christian knew, even before his visit to Israel, and before having any experience there, that he could lower the cost of drilling by half. I explained my views regarding the role of government in the oil industry.

When I noticed his close attention and receptiveness to my various arguments and heard his pertinent questions, I said: "Look, Israel. You are on the Board of Nafta. There is a custom in Nafta that, when they don't find oil, drilling to the projected depth, they decide to deepen the well." He agreed that this happened frequently in Nafta. I then explained the factors taken into consideration when a well is planned. Let us imagine, I said, that there were four prospective horizons. The first three were positioned in the first eight thousand feet of the projected well, and the fourth one "D" horizon, was three thousand feet deeper, at eleven thousand feet. If the "D" horizon appeared to be quite prospective, the well would be planned to go to eleven thousand feet, right from the start. A drilling rig suitable to drill to this larger depth would be ordered. If, on the other hand, the "D" horizon did not seem to justify the additional cost (usually double, as the cost of drilling rises in a geometrical progression with depth), the well would be designed for eight thousand feet only. This decision dictated the planning of the casing program, the drill bit size, etc. A rig for an eight thousand foot depth capacity would then be hired.

"Let us then examine the logic of a decision to deepen this eight thousand foot well to the "D" horizon, three thousand feet deeper, if oil was not found in the first three zones. Let us see what the economic ramifications of such a decision are," I said.

The fact that oil was not found in zones A, B and C in the first eight thousand feet, which were considered to be the most prospective, should obviously discourage one from going deeper, to the least prospective zone. "Instead you, in Nafta, drill an additional three thousand feet, with an unsuitable drilling rig, which is hardly capable of going to this additional depth. It more than doubles the cost of drilling the well. Such deepening, with an unsuitable rig, usually also results with "fishing" operations which increase the cost even more."

At the end of this meeting, which lasted over four hours, Dickenstein expressed his deep appreciation for my coming to see him. He said he had learned more in this meeting than he could have learned in a month-long seminar.

Several weeks later I heard about a rumor going around Nafta's Board that "Zvi Alexander succeeded in pouring poison into Israel Dickenstein's ear."

Ziqlag "2" went deeper than Ziqlag "1." It was again drilled in record time and at comparatively very low cost. Mayflower again sent its three experts, but with a new black book. There were no fishes and no fishing operations. Everything went smoothly and fast. But no oil or gas was discovered except non-commercial gas shows. The seismic picture we had obtained in our last survey of the Ziqlag "2" area was completely misleading.

Seismic surveys are based on creating strong sound waves on the ground, either by detonating explosives or thumping the surface with very powerful thumpers. The resulting sound waves travel through the earth's crust and are reflected each time a different layer of rock is encountered. These reflections are then plotted and the subsurface contours of different horizons put on a map. What one looks for is closed structures which may have acted as traps for oil or gas, if in fact oil and gas were generated in the area. This, in a nutshell, is a layman description of seismic surveys—the main tool of oil exploration.

The problem in Israel is that a thick limestone rock layer from the Cenomanian Age underlies most of the coastal plain areas of Israel. This coastal area happens to be the most prospective area in Israel from the point of view of oil exploration. The limestone layer is very cavernous and full of water. It therefore absorbs most of the sound energy generated in the seismic survey. The reflections from the layers below the limestone are thus weak and inaccurate. This is what happened to us in Ziqlag "2." The structure that we saw on the seismic maps did not exist at all.

* * *

I.N.O.C. also had two valid licenses in Caesarea and Netanya, where a lot of seismic surveys had been done in the past. We developed internally some exploration ideas relating to these areas and Mayflower agreed to join us again under the same formula as before: Mayflower 50 percent, Paz 25 percent, and I.N.O.C. 25 percent. I had agreed with Ray Christian that the Netanya joint venture would recognize a substantial part of the past investments in seismic surveys, in the relevant areas, amounting to several hundred thousand dollars. This would free I.N.O.C. from any cash requirements needed for drilling the Netanya wells and even give us some cash income to support our daily existence.

Ray Christian, Barth Walker (the chairman of Mayflower and their legal advisor) and Walter Neustadt, a wealthy oilman from Ardmore, Oklahoma, who was in Mayflower's group of investors in the Israeli venture, arrived in February 1968 to conclude the Netanya agreement.

Adi Ephrat, our legal advisor and I picked them up from the airport and brought them to the Tel Aviv Hilton.

When we sat down to chat, Ray Christian mentioned that he stopped by at Belco's offices in New York, on the way from Oklahoma. He had met with Robert Belfer, Arthur Belfer's son, who was now second in command in Belco, to discuss the possibility of Belco's joining Mayflower in Netanya and providing the bulk of the funds needed for drilling the Netanya wells. Ray Christian then started to mumble and stutter and became a little red in the face. He said that Robert Belfer objected to the agreement Ray had made with me regarding the payment to I.N.O.C. for past seismic surveys. Furthermore, Belfer insisted that Belco, rather than I.N.O.C., be the joint operator with Mayflower. When Ray Christian told him that he had agreed to recognize I.N.O.C.'s former seismic expenses, Belfer suggested a telephone call to Minister Pinhas Sapir, which would solve all the problems with I.N.O.C. Robert Belfer made a snuffing out gesture: Sapir will do "Fuu" and the problem will disappear. I then understood a strange telephone call I had received a couple of days earlier at 1 a.m. from Robert Belfer in New York. He was bubbling with enthusiasm and said: "Zvi, we are joining Mayflower, we are going to drill in Netanya, we will find oil for you, etc." I did not understand exactly how Belco fitted into the picture, but as Ray Christian was on his way I did not bother about it.

When I heard Ray's story, I had the feeling that I.N.O.C. was slipping from my hands. I knew that if we would not receive the payment for the past seismic we could not survive. We would not have the money to pay our share of the Netanya venture. I knew that if I lost the operatorship, we would not receive payment for our personnel and we would lose our standing completely. This would prevent us from operating in the future. All this went through my head whilst Ray was talking.

When Ray Christian finished, I made an instantaneous decision and told him that if Mayflower would not honor the original agreement, he and his colleagues should take the first plane out, back to Oklahoma. He asked me whether I was serious, and I replied that I was dead serious. Ray asked for time to consult with his colleagues. They all walked out to another room, and Adi and I remained in the first room. Adi was white as a sheet, and I probably did not look much better. Adi said that I was putting the whole future of I.N.O.C. on one card. I replied that if we accepted Belfer's terms, we were dead anyhow. Better to go down with the ship, standing up, than dying a slow death.

A full hour passed. Then they all came back and Ray said: "You win. The agreement stands as we agreed originally."

I said: "Fine. What time shall we meet in the morning?"

He replied: "No, you come now to my room. I am going to call Robert Belfer."

Ray called Robert Belfer in New York and told him that he related their New York conversation to me and that I then told him to pack and take the first plane back to Oklahoma. There was a long tirade at the other end of the line, and I heard Christian replying to Belfer (who must have claimed that I was bluffing and that if Christian had been firm I would have given in): "Maybe Zvi was bluffing, but I did not want to take the risk and go back to Oklahoma." Again there was a long tirade from Belfer, in which Belfer must have used some strong language, as I heard Christian saying: "Bob, we are not used to this kind of language in Oklahoma. Either you apologize or I hang up." This exchange was repeated several times over; finally Belfer apologized and the conversation was over.

We met the next day and signed the agreement as originally conceived. The Netanya venture was ready to commence, and we were going to have the money which assured our survival. Throughout the years of my acquaintance with Ray Christian we never mentioned the Belfer incident again. We drilled two Netanya holes. Both were dry. Par for the course.

I heard again from Bob Belfer two years later. He called me up and said that he heard and saw that I was exploring all over the world, and why did I not invite Belco to join. I do not recall what I replied. I have not seen or called the Belfers again since that incident that happened so many years ago. In retrospect, I may have judged Bob Belfer's behavior too harshly. It is quite conceivable that he did not realize how important this deal was to us and that our very existence depended on it. All I remember was that I had helped them greatly when they came to explore offshore Israel, and they knew it and appreciated my help; but here they wanted "to rob the poor man's lamb." They wanted to rob me of the only remaining asset I had.

Chapter Four

First Steps in Black Africa—Ethiopia

I have been asked many times why God has "punished" his chosen people, the Jewish nation, by not blessing them with oil. Why are there immense oil reserves in all the neighboring countries whilst Israel remains "dry"?

My reply is that it is not true that there is a lot of oil around us. There is no oil in Lebanon, Jordan, or Syria, except the north-eastern corner of Syria, which is the north-western extension of the Persian Gulf rift. There is also, despite the common belief, no oil in Sinai. What we in Israel call "Sinai Oil" is in fact the oil in the Gulf of Suez. The rich oil deposits of the Gulf of Suez are in a graben, an elongated geological depression, narrow and long, filled up with marine sediments in the course of many millions of years, which created the oil deposits. Many grabens around the world are oil rich provinces.

Furthermore, the whole Mediterranean Basin is oil-poor. There are no large oil deposits in Lebanon, Syria, Turkey, Greece, Italy, southern France, Spain, Morocco and Tunisia. The only oil-rich areas in the Mediterranean Basin are in Libya, Algeria and the western desert of Egypt, which are in a different geological setting. One more area that has oil and very large quantities of gas is the Nile Delta, both onshore and especially offshore. It is quite possible that the substantial quantities of gas that have recently been discovered in deep waters, offshore Israel (see chapter 25) originate in Nile river sands, which travelled northwards in the direction of Gaza and Israel.

It is possible that in the geological past of Israel, many million years ago, there were substantial oil deposits which accumulated in the very large structures in the southern part of Israel, the Negev, the Ramon Circ, the Big Circ and the Small Circ, and other large structures. The problem is that the Afro-Syrian fault, which runs along the Dead Sea and the Arava depression, is quite young, approximately half a million years old. This fault may have caused all the oil accumulated in those large structures in the Negev to leak out and evaporate. The large asphalt blocks that from time to time float up in the Dead Sea may be indications of this phenomenon. By the way, the Dead Sea rift is also a graben. The problem is that it is extremely narrow, very deep, and highly fractured. Therefore it is quite unlikely that commercial quantities of oil remained there. Many wells were drilled there, but with no success.

* * *

I slowly started coming to the conclusion that we would not be able to solve the country's energy problems by drilling only in Israel. The huge structures in the Negev, in the southern part of the country, were all drilled and found to be dry. The Judean mountains, then under Jordanian occupation, which geologically were one big structure, were drilled by John Mecom from Houston and found to be dry. With each new wildcat resulting in an additional dry hole the hopes of hitting the big bonanza started to dim. I came to the conclusion that Israel must go outside its borders to find oil.

It was not easy to come to this conclusion, and it was even more difficult to convince others that, painful as it was, it was the only logical answer. It should be remembered that oil had already been discovered in Israel, and the small quantities just encouraged the hope that large accumulations of the black liquid were hidden somewhere—in the Negev, near the Dead Sea, in the plains or in the Galilee. No less important, Israel is, after all, a Middle Eastern state, and the largest and richest oil reserves are found in the Middle East. Why should we have been shortchanged?

This and more. It is true that the Arab boycott did not succeed in strangling the Israeli economy, but it did cause significant harm, particularly in the area of oil supply. Terence Prittie and Walter Henry Nelson, in their book *The Economic War Against the Jews* (Secker & Warburg, London, 1977), point out that, already in 1950, Egypt, the leader of the Arab League, declared that oil is "strategic merchandise," and that the Arabs must make every effort to prevent its delivery to Israel. Prittie and Nelson write: "It is a war which knows no boundaries, a war of economic coercion, extortion and blackmail; it involves boycott and blockade, and is waged against the world business community, using wealth derived from Middle East oil as its principal weapon...."

This is a pretty accurate description of the Arab boycott, particularly in the first two decades of the State of Israel's existence. Although Israel's political/security leadership managed to find sources to provide her with oil, the Damocles' sword of the Arab boycott still hung over Israel's head constantly. Israel's oil suppliers had to work in secret and did everything to camouflage their terrible "sin": the provision of oil to the boycotted state. In the heart of those responsible for Israel's oil supply there was always awareness of its insecurity. This apprehension was indeed justified: For years Israel received a steady stream of oil from Shah-ruled Iran; then came the Khomeini revolution in 1979, and in an instant Israel was severed from its main source of supply. Fortunately, by that time there were already many states which produced oil in large quantities and which ignored the edicts of the Arab Boycott Bureau. Ironically, one of them was Egypt—the initiator of the boycott—which by 1979 had signed a peace treaty with Israel, under the terms of which it would, *inter alia,* enable Israel to acquire oil from its production.

In the 1950s and 60s, however, Israel's economy was still in its infancy, and the Arab boycott was very powerful and threatening indeed. Israel's dream was to discover oil in quantities large enough for the state's needs. No one dared to hope that Israel would turn into a Saudi Arabia or a Kuwait. All it longed for was not to have to import oil. This was a relatively modest dream and, on the face of it, realistic. Had we not already discovered oil in Heletz? Therefore, let us go on, somehow, with the search for oil until we find the place where the greatest accumulation is hidden; this would free us from our dependence on foreign sources.

When I set out to destroy this dream, I was on my own. Very few were willing to listen to me. No one likes the wrathful prophet who wields the hammer on dreams and hopes. Furthermore, my position endangered the protected interests of powerful individuals and bodies—the oil exploration companies and their staff, geologists, politicians and senior officials. But I was certain that I was right and decided to be adamant. I had to try to convince others that we must direct our efforts outward, beyond our borders.

I remember some of the arguments I presented to Dr. Arnon, the director general of the Ministry of Finance, when I rode with him in his car from Jerusalem. I suggested that Israel should create a task force to examine favorable prospective areas around the world for oil exploration. They would not necessarily have to start roaming around the world for that purpose. Much in the way of such studies and review can be done in the office by going over published geological material.

I told Dr. Arnon that Israel's practice of trying to join in a foreign exploration effort—on the rare occasions that such an opportunity was offered to us—was completely wrong. We used to create an *ad hoc* committee, which had to include at least a dozen people, in order not to offend anybody. By the time they all convened and finished arguing, the opportunity had vanished. Not that we lost much. I told Dr. Arnon that, in my view, the opportunities presented to us were the worst "dogs" in the industry. If a project has any real merit, there is enough money in the world willing and able to finance it. If somebody bothers to come to Israel to ask for our participation, the project's chances are probably so poor that he could not find funds elsewhere.

Later, when I joined I.N.O.C., I developed a list of arguments supporting my belief that Israel's energy requirements can only be found abroad. I said that many countries came to the same conclusion and created the tools necessary to pursue such a policy, and that if we did not consider all the non-Jews to be stupid, we should perhaps follow in their footsteps. Moreover, there were many different ways of joining this game. It was not necessary to go everywhere with the Israeli flag flying. We did not have to brand every drilling rig and every geologist with the Star of David.

I remember one argument I used in many discussions, that I later voiced with various ministers, members of the Ministerial Economic Committee. I used the example of France, which is fifty times the size of Israel and was exploring in over one hundred countries around the world. The geological setting of France was not very different from that of Israel. They had some oil and gas, not very much, and mediocre oil prospects. If the French had followed the Israeli pattern, they would have had hundreds of drilling rigs punching holes in France, and the map of France would eventually look like Swiss cheese. Instead, they came to the conclusion to support world-wide exploration efforts, made by both private and government French enterprises. And France was not a unique case. Other countries followed similar policies.

I also emphasized that France makes all this tremendous effort abroad in spite of the fact that it can buy cheaply all the oil it needs from the Arabs. The French do not suffer from the Arab boycott. Furthermore, they have other sources of energy, coal and atomic energy. In spite of all this, they explore from the Arctic to the Antarctic and from Indonesia to America.

Another argument I used in all these various discussions was that the oil rights of companies around the world are, surprisingly, very stable and very secure. Even Muammar al-Gadafi had allowed the oil companies in Lybia, which had obtained their concessions under King Idris, to continue operations undisturbed. Egypt, in spite of the Six-Day-War, and Nasser's claim that it was the Americans, and not the Israeli forces, that destroyed the Egyptian air force, allowed the American oil companies to operate unhampered. Nasser severed Egypt's diplomatic relations with the US, but the American oil companies continued to pump, and explore for, oil in Egypt. It is interesting that this phenomenon is not widely realized, even among oil companies. Oil concessions are very stable.

There is a reason for this situation. Oil exploration is a very long-term affair. If you are lucky, and you find oil, after some years of geological studies and seismic surveys, you will start producing and selling it only several years thereafter. The development of the field, building of storage tanks, pipelines, etc., will take a great deal of time. In other words, in the best-case scenario, one will start producing seven to ten years after the granting of the license. The company then has a twenty to twenty-five year production license, during which it will be able to recoup its investments and make a profit. Even the most stupid political regime can understand this economic reality. Therefore most of the countries in the world leave the oil explorer undisturbed, whatever changes occur in the government and in the regime.

I started to present the case for foreign exploration to Dr. Dinstein in late 1967, soon after his return to the Ministry of Finance. Israel, then, was already producing oil from the captured Sinai oil fields. Still, we did not know how long we would be able to hold on to these fields, before their eventual return to Egypt. Furthermore, the Sinai oil fields supplied only one third of Israel's consumption.

Dinstein said: "Look. I won't be able to give you any money. Perhaps the way to go about it would be to persuade Delek and Paz to undertake a majority interest in this venture. I might then be able to allocate some funds to I.N.O.C. to participate with them."

Delek and Paz were marketing oil and oil products in Israel. They were, and are to this day, a kind of monopoly, operating under a "cost plus" system. They therefore had a "License to Print Money."

I first went to see Emanuel Racine, the head of Delek. He liked the idea very much but said that I should first talk to Paz. Paz was the leader of the oil marketing companies. Wherever Paz goes Delek follows. I then went to see Meir Sherman, the head of Paz, and our partner in the Ziqlag venture.

Sherman wrote a short letter to Max Fisher in Detroit, asking for his approval. Fisher replied in one sentence: "Have you found all the oil in Israel yet?" This was the end of the story. The same Max Fisher offered to finance, from his own pocket, all I.N.O.C.'s foreign exploration, for a half interest in our concessions. But this was two years later, when we were already prospecting in several African countries.

Max Fisher, one of the leaders of the Jewish community in the US and one of the wealthiest Jews in America, represented Signal Oil and Gas on the Board of Directors of Paz. Already in 1954 Signal started secretly supplying oil to Israel from its participation in the Iranian consortium. Following this supply contract, Signal acquired one-third interest in Paz. Signal's one-third ownership of Paz was a closely held state secret. To all intents and purposes, Max Fisher acted as a majority owner of Paz, even though he did not own one single share in it.

Dr. Dinstein fully realized the importance of oil exploration abroad. In spite of the fact that he could not allocate any money for this effort, he passed a resolution in the powerful Ministerial Committee for Economic Affairs authorizing I.N.O.C. to be the "Government's Arm in Obtaining Oil and Mineral Exploration Rights Abroad." Nobody remembered to rescind the resolution of shutting down I.N.O.C., and it is probably still in force. Also, none of the ministers asked where the money for these rights and exploration effort was going to come from.

And so one day, in late 1967 or early 1968, on the evening news, I heard the announcement of the ministers' decision authorizing our foreign exploration effort.

We started with Ethiopia. There were several reasons for this. With the conquest of the Sinai peninsula in the Six-Day-War, Ethiopia became a close neighbor. Ethiopia guarded the entrance to the Red Sea at Bab El Mandeb. It had free areas for exploration offshore, which had been relinquished by Mobil Oil, the second largest oil company in the world. Rumor had it that Mobil had a gas blow-out during its drilling offshore. It relinquished the area only in the aftermath of the Six-Day-War and the resultant closure of the Suez Canal. The economic, political and military ties between Ethiopia and Israel were at their zenith. And, last but not least, the Israeli ambassador to Ethiopia was Uri Lubrani, a close personal friend of mine, who had an excellent standing in that country.

Ethiopia is very different from other African countries. Most of the country is situated on a very high plateau, almost ten thousand feet high. This is probably the reason why Ethiopia remained independent throughout the thousands of years of its recorded history, except for the short period of Italian occupation. Prior to World War II, Mussolini wanted to show Italian prowess and waged a modern war against the Ethiopians, some of whom still wielded a bow and arrow.

The isolation from the rest of black Africa created by the inaccessible plateau was responsible for the development of a different race there. Their body and facial features are completely different from those of other African people. They have thin noses and thin lips. Both men and women are very handsome, and some of the women are real beauties. They all have a proud bearing and move very gracefully.

We found the country to be extremely poor. On the roads leading from Addis Ababa, the capital of Ethiopia, we often saw vendors offering a single chicken. Many children stood all day long offering two or three eggs. Wherever we stopped, we were immediately surrounded by poverty-stricken children offering various services. One offered to show us crocodiles sunning themselves on the river bank, another wanted to show us monkeys, etc. All were trying desperately to earn some money.

In the other countries of Central and West Africa, the vegetation was lush and everything grew abundantly. One could imagine that people did not have to work to eat. They could easily pick fruits from the trees. They did not need any clothing to speak of to protect themselves from the elements. Not so in Ethiopia. Though the weather was beautiful, cool during the day and quite cold at night, there was no lush vegetation. That was the reason for the great poverty.

Ethiopian ties with the land and the people of Israel went back thousands of years. The biblical story of the marriage of King Solomon and the Ethiopian Queen of Sheba is to Ethiopians a living part of their history. The modern ties with Israel developed after the establishment of the State of Israel. When I first visited there in late 1967, those ties were at their full bloom. There were Israeli advisers in cotton-growing and other agricultural fields. They were mostly members of Kibbutzim (communal settlements in Israel) who were used to rough conditions and, unlike other foreign experts, preferred to work in the field rather than behind a desk. Israel also supplied instructors to various branches of the Ethiopian Army and Air Force. There were Israeli advisers with the Ethiopian intelligence services. When the taxi driver in town heard that we were from Israel, he said: "Israel very strong, very strong people."

The number of Israelis stationed in Ethopia was unusually large. One of the officials in the embassy was leaving after three years of service. We must have attended three or four farewell parties in his honor during one of our visits.

In one of our trips we stopped at a roadside café, hundreds of miles from Addis Ababa. It was very strange to suddenly hear Hebrew spoken there. It was the wife of an Israeli Air Force instructor, admonishing her children who were playing noisily.

My wife joined me on one of my trips to Ethiopia. When we landed, she was advised to walk slowly because of the high altitude and the resultant lack of oxygen. She took several sight-seeing trips while we were negotiating and she always came back with interesting stories. One I remember in particular. She was told that there are different types of marriage contracts in Ethiopia. There is a contract for a month, for a year, for two years, and so on. Quite an interesting arrangement. You do not have to stay married to one woman for the rest of your life.

We decided one evening to go, by ourselves, to a native Ethiopian restaurant to taste the local fare. There were no plates and no cutlery. A very large pita bread, of a spongy texture, was served, covering the whole top of a small serving table. The food, which was extremely hot, was put directly on the pita. A slice of the pita was then torn off, and with it we picked up morsels of food. It was a very interesting experience.

There was a plague of flies in the hotel where we stayed, which the hotel staff could not get rid off. As this was the Addis Ababa Hilton, they surely tried hard. When my wife inquired about the reason for so many flies, she was told that the emperor's stables were located nearby and there was no way to eradicate all the insects.

The thing that struck my wife the most during this trip was that after several days, when a blond European woman walked into the hotel lobby with several of her children, she seemed pale and anemic in comparison with the majestic appearance of the Ethiopian women. It appears that the appreciation of beauty is an acquired taste which changes in different circumstances.

My first trip to Ethiopia was in late 1967, even before the official authorization for I.N.O.C. to explore abroad. I met with the minister of mines, Major Asefa Lema, and his deputy, Ato Teshome. The minister was an outstanding individual, a highly intelligent, dedicated and completely honest man. When he was nominated minister, he taught himself what the oil business is all about and knew what questions to ask and what particulars to insist upon.

I often wondered why, in Africa, some of the ministers made the effort to study thoroughly the field they were appointed to supervise, whereas in Israel I do not remember many ministers who bothered to do the same. I have never found an answer to this question.

The deputy minister, Ato Teshome, had a different personality altogether. He was much younger, a lawyer, and a graduate of McGill University in Canada. The minister did not appreciate him very much. All the preliminary negotiations were made with the deputy minister, and then we had to come back to the minister for a repeat performance. He did not really trust his deputy's judgment.

We found the Ethiopians extremely suspicious. Any word innocently uttered might start a chain of suspicions which could lead to a break in negotiations. "You have said this, then you are not serious about your obligations here," and so on and so forth. One had to be extremely cautious in one's statements and behavior.

One day our lawyer, the late Micki Deouell, complained to the minister that we were going over every clause twice, first with Teshome and then the minister. He was immediately thrown out of the minister's office. Only our ambassador's intervention brought the situation back to normal.

Uri Lubrani, the Israeli ambassador helped me immensely. His prestige was very high in Ethiopia, the country's relations with Israel were excellent, and his shrewd advice and intervention, whenever requested, were of great help.

Several times there was a week-long pause in the negotiations. As it was difficult to return to Israel, I went several times to Kenya to wait for the resumption of the negotiations. There were only two possible flights a week from Addis Ababa to Athens. To fly from Israel to Ethiopia, one had to go via Athens and to overfly Egypt. When the airplane left Egyptian airspace, we always expressed a sigh of relief.

On one of those trips to Kenya we had an amusing incident. We were sitting in the office of the Israeli ambassador to Nairobi, discussing the possibilities of exploration in Kenya. The ambassador was a former Labor Union official. His economic attaché had even better credentials. He was in charge of the drivers at the Labor Union headquarters in Tel Aviv. During the discussions, the ambassador asked me why we were interested in exploring abroad. I asked him what his security clearance was. He said it was "the highest." I insisted that I must make sure, as I was going to tell him a state secret. Then I whispered: "We do not have enough oil in Israel." He got very red in the face and did not ask any more silly questions.

One of the serious problems we encountered in Ethiopia was the impossibility to show the minister our past record in oil finding, as well as a proper balance sheet. We had somehow overcome this obstacle with the help of Ambassador Uri Lubrani, and that of Dr. Dinstein when he visited Ethiopia in February, 1968. Dinstein had an audience with the emperor, Haile Selassie, and met with the prime minister, the minister of mines, etc. His visit was so prestigious and successful that it enhanced our chances considerably.

During Dinstein's visit we had a long discussion following a dinner party at Lubrani's residence. We discussed my future plans for I.N.O.C. I told them that I planned to raise money in the international market place, on Wall Street. Dinstein said that hair would grow on the palm of his hand before that happened. Luckily he was wrong.

We also applied, in Ethiopia, for large areas for mineral exploration. We ended up with exploration permits for gold, uranium and nickel in different parts of the country. All in all, we took upon ourselves obligations for millions of dollars of exploration work, when we had hardly enough money to pay for airline tickets. The mineral activities we could drop if we did not find partners to fund the exploration effort, but the oil exploration obligations were for real. We had to perform.

One of the reasons why I wanted to acquire as many concessions as possible, either oil or minerals, was my belief that the larger the portfolio of assets we would be able to show, the better the chances of raising money in America were going to be. Later events proved that I was quite correct in my conception.

Years later when the state comptroller was sitting in our office reviewing all our past activities, he had two "serious complaints." How come I took upon myself so many millions of dollars of obligations without prior approval by my Board of Directors? The other complaint was that mine was the only signature on the concession agreements, whereas the state regulations stipulate two signatures for important documents. My reply was that there was no point in asking Board approval in advance, for an expenditure of millions, when there were only a couple of thousands in the bank account. As to his second complaint, I replied that I was not going to waste our meager resources on an additional, unnecessary airline ticket.

At the end of a month-long review by the state comptroller of all contracts, correspondence and other relevant documents in our offices, the comptroller told me in our final meeting, "In spite of all the infractions and unauthorized activities that you have undertaken, I shall pray to God that there would be more Israelis like you." I was very proud to hear this statement.

We had very little competition in Ethiopia. Later in the year a Canadian group appeared. It included the former British Governor of Aden, who had many important contacts in Ethiopia. They applied for the same offshore areas. By that time our negotiations were well advanced and they could not dislodge us any more.

We signed the concession agreements in Ethiopia in late 1968. While we were negotiating in Ethiopia, an American lawyer by the name of Herbert Glaser visited Israel. He was one of the directors of a fabulously successful insurance company in California, Equity Funding Corporation. They were the envy of the insurance industry in the United States. Barth Walker, the chairman of Mayflower, was also the chairman of a small Oklahoma insurance company. He told me that he would pursue every avenue to copy the successful formula of Equity Funding. Barth was lucky that he did not discover the formula. It led to a fifteen-year jail sentence for the Equity Funding president. But this happened five years later. In 1968, Equity Funding was flying very high.

Herbert Glaser was very interested in our oil project in Ethiopia. He said that they wanted to diversify and that energy was a suitable field for them. Immediately after we concluded our agreement with Ethiopia, we signed up with Equity Funding. They took a 25 percent interest and paid double, i.e., for 50 percent, this being the terms of the "farm out." In addition, they reimbursed the costs and expenses we had incurred in the Ethiopian project. They paid the cost of a senior geologist, as well as the overhead and operators fees we charged. All in all, it was a very substantial contribution to our finances, but less than peanuts for them.

We hired Efraim Aharoni, who was a senior geologist in Nafta. He had one of the best geological brains I have ever encountered. Again there was an outcry in Israel that we were stealing people from other oil companies. But, in general, they left us alone. We virtually stopped exploring in Israel. The danger of foreign investors in Israel thus disappeared. We did not bother anybody. We did not ask for money. There was some envy of our constant travel abroad, but it did not affect us very much.

There was one plus in not receiving a budget. We did not have to ask permission to go abroad or request foreign currency allocations. Some of the money we received from our joint venture partners we kept abroad. With these funds we paid for our travel and other foreign currency expenditures.

Several years later we realized that we had ten foreign subsidiaries registered in various places around the world. As registering a foreign subsidiary required a lengthy procedure of approval by the Economic Ministers Committee, we did not bother with it. I do not even think we knew that we were not complying with the law of the land. When we found this out, we asked for and received retroactive approval for the establishment of all of our foreign subsidiaries. In any case our Board of Directors knew about and approved each and every step we took.

By that time, we already employed two separate law firms to handle our affairs. Adi Ephrat's firm, Goddard and Ephrat, handled our corporate affairs and farm out negotiations, whereas Micki Deouell's firm, Levin and Deouell, looked after our concession negotiations.

We also built up a prestigious Board of Directors. The Ministry of Finance nominated some top officials, the Foreign Office added some ambassadors. The head of the Civil Service nominated himself to the Board and the Ministry of Development named the petroleum commissioner. It was a pleasure working with this Board. I always reported to them the full extent of our negotiations and obligations. They must have somehow believed that I would eventually find a way out. How I dared to take on these enormous obligations for I.N.O.C. baffles me to this day.

We had excellent relations with the Ministry for Foreign Affairs. We loved them and they loved us. We supplied them with real life activities instead of barren diplomatic contacts. They provided us with current information and gave us all the help we asked for. We were permitted to use their cable and telex network, which was fast, efficient and confidential. All our communications with our men abroad went through this network.

One gift that we received from the Ministry for Foreign Affairs was worth more than its weight in gold. This was the late Hanan Yavor, the former Israeli ambassador to Ghana, and subsequently Nigeria. Hanan was a devoted director of I.N.O.C., friend, confidant, and my greatest supporter. Jewish folklore states that at any given time there are thirty-six righteous men in the world. Without them, the world could not survive. Hanan Yavor was definitely one of them.

One of the most prominent leaders of Nigeria once told me that Hanan Yavor was not just the Israeli ambassador to Nigeria, but the Nigerian representative in Israel. One of the things that impressed the Nigerians most was Hanan's custom to go to the refrigerator and help himself, when he visited a Nigerian but did not find him at home. This showed a degree of trust and friendship that they so much needed during the first years of their independence.

Hanan Yavor travelled with me to Ethiopia. In Israel, he participated in all the meetings I had with Dr. Dinstein and other officials of the Ministry of Finance as well as in meetings with the Ministry for Foreign Affairs. All this he did without any remuneration. He loved our work and considered it essential for the welfare of the state. For him, it was a just and adequate reward for his toils. He was a true Zionist.

* * *

In late summer 1967 Jack Grynberg, the oilman from Denver, came to Israel. He had come to the conclusion that there is a geological similarity between the area south of the Dead Sea in Israel and the large El Morgan oil field in the Gulf of Suez. For this reason he obtained an oil concession in the area south of the Dead Sea. He succeeded in interesting US Refining and Smelting in his ideas. They took a 50 percent participation. With this interest in hand, he approached Max Fisher in Detroit who took 25 percent of the deal. Paz followed suit and took the other 25 percent.

Grynberg was a complete outsider on the Israeli scene. He had left Israel at the age of seventeen, in 1947, and his first return visit came twenty years later, in the mid-1960s. For him to succeed in selling an oil exploration deal in Israel to Max Fisher, one of the most influential and pro-Israel American Jews, and to Paz, the largest and the richest oil company in Israel, was a far greater achievement than selling refrigerators to the Eskimos.

Jack Grynberg's "farm out" to Fisher and Paz strengthened my belief that, once you obtain prospective acreage and you have it under license, and once you create a good geological story, you stand an excellent chance of finding partners. These partners would usually pay you a premium for your past expenditures and from then on fund the exploration effort.

* * *

Since the mid-1950s Israel has helped many developing countries around the world. These ties, mostly economic and sometimes even military were most pronounced in Africa. There were many reasons for this development. The African countries, recently liberated from the colonial yoke, were suspicious of their former masters. Israel did not carry this stigma of colonialism. The Israelis were also not strictly white, which was another plus in Africa. The fact that Israel was a young country which, though not blessed with natural resources, had succeeded in a short time to turn the desert into a blooming garden, was much admired by the newly-independent countries. When they looked at Europe or the United States, they saw there the result of hundreds of years of development. They did not want to wait so long.

Israel made a tremendous and costly effort to help Africa, because it tried desperately to find friends to balance out the Arab hatred surrounding it from all sides. The way to achieve it, at least partially, was to find friendly countries situated outside the belt of hatred of Arab countries towards Israel.

Another reason for Israel's great success in the developing world was the fact that its experts loved to work in the field together with the local population. This was in complete contrast to other foreign experts, who conducted their aid programs from behind desks in air conditioned offices. Israel conducted various courses and seminars for African and other developing countries' students in specially built schools in Israel. I remember several Ethiopians I later met in Addis Ababa who spoke Hebrew fluently. Some of their children who were born in Israel even had Hebrew names.

I.N.O.C.'s activity in Africa began in East Africa, in Ethiopia as described previously. In the next stage we decided to "attack" West Africa, which appeared to be very promising. Very substantial reserves of oil and gas were discovered both in Nigeria and Gabon. The other countries in West Africa, which were still "virgins" where oil exploration was concerned, waited impatiently for the drilling bit. The two countries that interested us most in the beginning were Ghana and Ivory Coast with which Israel maintained excellent relations.

As a first step, I cabled the ambassador and asked him to obtain all the information regarding an application for licenses in Ghana. The ambassador met with the minister of mines and was told that he would not object to our application. The ambassador was also advised that the minister would be in London on December 15. As I was travelling to the States, I stopped off in London and met with J.V.L. Philips, the minister. He told me that if we had the expertise and the necessary funding, we could submit an application.

This must be done within the next six weeks, before January 31, 1968, as the list of applicants would then be closed. He informed me that they had not yet enacted the necessary laws and regulations, but that they were in the process of drafting them. In the meantime, they wanted to have the complete list of applicants to be able to check their credentials. If we passed the test of suitability, he would welcome us in Ghana.

When I returned to Israel and considered our chances in Ghana, I came to the conclusion that we might be disqualified by the Ghanaians for lack of expertise in offshore exploration, and also for lack of financial resources. Furthermore, I was concerned that on the Israeli side somebody might wake up and say; "Hey, what are you doing offshore Ghana. You are going to embarrass us all and spoil our good relations there." I realized that we would need a partner who could demonstrate both in Israel and in Ghana that our group possesses sufficient technical expertise to conduct the exploration activities. I therefore called Ray Christian and asked him if he wanted to explore in Ghana. "Where the hell is it?" he asked. "In Africa," I replied.

"Is it any good?"

"I think so," I replied.

"All right, I will go," he said.

With Mayflower as a partner, we were in a different league. There would be no more questions in Israel as to whether we were doing the right thing, after Mayflower's exceptional performance in Israel. Also, there would be no questions in Ghana relating to our technical competence.

In talking to the Ghanaians we used the Avis Rent-a-Car approach. We are number two; therefore we have to try harder. Our argument in Ghana, which we later adapted to other places around the world, was that we would "do our darnedest to find oil soonest" and produce it as fast as possible. We argued that the major oil companies had a surplus of oil. They had global interests which might dictate to them to produce oil from politically unstable areas first and leave the oil in the ground in Ghana. We, on the other hand, were hungry. We would run faster. "Small is beautiful." The argument worked.

The fact of the matter is that this argument happens to be true. Ghana gave licenses to fourteen different groups, which included major oil companies, large independents, and small companies like ourselves, Grynberg, etc. Ghana's neighbor to the west, Republic of Ivory Coast, gave all the offshore to a consortium of majors. As a result, there were three times as many holes drilled in Ghana than in the Ivory Coast.

I suggested to Ray Christian that we include Marvin Billet and John Christensen (Chris), the prominent geologist from Rome who helped us with our Ziqlag project, in our group. Marvin would assist us in finding farm out partners, while Christensen would furnish a world-wide geological approach, and prepare the geological reports for future investors. Chris, I thought, with his distinguished looks, demeanor and reputation, would also add strength to our group vis-à-vis the Ghanaians. I asked Ray to invite them to join in. They would have to pay their appropriate share of the expenses. Ray called me back and said they would love to join, but Billet refused to pay expenses.

"They are out," I said.

Ray called again, after a while: "They are ready to pay."

"They are in," I replied.

In March 1968 the negotiations in Ghana started. Ghana was quite a different kettle of fish. Whereas in Ethiopia we had almost no competition, everybody and his uncle was competing in Ghana. The West African offshore areas were one of the hottest prospective areas around the world.

The weather in Ghana, too, was completely different from that in Ethiopia. In Addis Ababa, the capital of Ethiopia, at the elevation of eight thousand feet, the climate is cool and pleasant. In West Africa, it is very humid and hot all year around. All you want to do is get out of the heat into the air-conditioned room in your hotel. The enervating climate had obviously influenced the local people's customs and behavior. They are very patient and have all the time in the world. Ghanaians are lovely people who smile and laugh a lot. Nobody is under pressure to accomplish anything. Punctuality is not necessarily considered a virtue. (Sylvan Amegashie, the Ghanaian minister of mines, who has remained my friend to this day, is always at least one hour late for a meeting. He walks in laughing, and declares, looking at his watch: "Central African time.")

An American lawyer we met in Ghana told us a story about an executive of an oil company who returned to his home office after ten days in Africa. When asked what he had accomplished, he replied that he had succeeded in making one telephone call. His superior thought that the man had lost his mind. After spending some time in West Africa, one can come to the conclusion that making one telephone call could be an important achievement.

* * *

When we arrived in Ghana, we found out that Sylvan Amegashie had replaced Phillips as the minister of mines. Amegashie was a certified public accountant, actually the first black C.P.A. in Ghana, and possibly in all of West Africa. He was educated in Britain and was head of the School of Management in Ghana before being nominated a minister by the military regime. A military coup had overthrown Kwame Nkrumah, the first president of Ghana, and the foremost leader of black Africa. They installed a government of technocrats, including Amegashie.

We saw and understood the deep disappointment of the people of Ghana with Nkrumah, who was at first considered a hero and a prophet. The only tall, modern building of Ghana had been erected by Nkrumah at great expense for one purpose only—as the venue for the meeting of the Organization of African States. The people were so incensed at this extravagance that they kept the building completely empty. It was called "Job 600." This number was probably the serial number in the Public Works department.

Sylvan Amegashie was an excellent choice for minister of mines. He was a businessman who understood that oil exploration was a business—albeit a very risky one—that would eventually make a profit or not survive. Many governments do not understand this simple truth to this day.

The government of Ghana prepared a first draft proposal for the new petroleum law and regulations. This draft was given to us upon our arrival. Each company was asked to submit its comments in writing, by a certain date. A new draft was then prepared by the ministry, based on the companies' comments. One day, while waiting to receive the second, revised draft, we came to the ministry and found Amegashie sitting at one end of a very long table and Prah, the permanent secretary, at the other end, with a pile of papers a foot high at his elbow. These were the various applicants' remarks to the proposed law. Amegashie shouted: "Clause Twelve." Prah called out, "Occidental Petroleum says this, Mobil says that," and so through the list of twenty or more applicants. I realized that the revised draft would not be finished for at least a month. I requested permission to leave Ghana. Amegashie said: "No, please wait. We will be ready in a couple of days."

I replied: "Mr. Commissioner (this was the title of ministers under the military regime), I promise you that I will be back in Accra within twenty-four hours of receiving your summons. In the meantime, I have a company to run, I also have to go to Ethiopia."

Amegashie reluctantly agreed. Six weeks later we received the new draft.

We made many comments on the new draft which we submitted by a certain date. At last, at the beginning of December, 1968, we received the final draft contract. This was professionally printed in a print shop. No more changes, no more remarks. "We heard you twice and we have made up our mind. Take it or leave it."

The final contract was tough but fair. It called for thousands of kilometers of seismic survey within the first eighteen months and for drilling at least twelve thousand feet of hole within the following year, an expenditure of many millions of dollars. The signing ceremony was scheduled for December 31.

In the middle of that year I realized that our original plan to obtain the license and then farm it out immediately to another oil company (the Jack Grynberg system, as we called it) would not work. Not only might we later lose the license, if we were successful in getting one, but we also might damage the good diplomatic relations between Israel and Ghana. We must show I.N.O.C.'s involvement in exploration activities from the beginning, and we would also have to see to it that all legal and work obligations be executed accurately and punctually. The problem was that by that time I got to know my partner Ray Christian very well. Ray had many outstanding attributes but being on time was not one of them.

I asked Ray to come to a meeting in London. We met at the Westbury Hotel, which at the time was the "oilmen's hotel." I explained to Ray that, although we had originally agreed that Mayflower would be the leader in our venture in Ghana, I.N.O.C. must now assume this responsibility. I told him that the venture was much more important than I had originally anticipated. We would have to fulfill our legal obligations, including payments of rentals, submission of reports, and so on, precisely on or before their due date. If we failed to do so, we might lose the license. We must also meet our work commitments on time. For all these reasons, we had to change the original agreement and I.N.O.C. would have to become the operator. Ray listened to me, did not say yea or nay, and we both returned home.

Back home, I wrote Ray a long letter listing all the reasons that I had given him during our meeting in London. I reiterated why we had to completely change the original agreement. I did not receive a reply. When, seven months later, Ray came to Israel to sign the operating agreement for Ghana with us, he not only remembered and agreed to all the paragraphs itemized in my letter to him, but also agreed to any and all implications resulting from these changes. He did not say, "Well yes, but I understood it in a different way," or other similar excuses.

During these negotiations Adi Ephrat, our lawyer, tried to improve I.N.O.C.'s position even more. From time to time he would say, "Ray, the Israeli government cannot agree to this, it cannot agree to that" and so on. Finally Ray got tired and said: "Adi Ephrat, to hell with you and to hell with your government. I have a written agreement with Zvi whereby we are the leaders. I understood Zvi's worries about fulfilling the obligations to the Ghanaian government on time and agreed to all the changes. If you dare mention the government of Israel again, we are going back to the original agreement by which Mayflower is the leader." The government of Israel was not mentioned again.

I was concerned that Ray's habit of delaying and procrastinating may cause us damage in Ghana once we started operating there. One meeting we had during the course of that year made me really worried. We had to resolve some problems regarding our future operations and suggested that we meet and find solutions. Ray suggested I come to Oklahoma, I suggested he come to Tel Aviv; we finally settled on London.

The meeting was scheduled for mid-October in 1968 when there was school vacation in Israel, due to the Tabernacles holiday, which meant that my wife could join me. We arrived at the London Hilton. Ray was obviously late one day, when he arrived with his girlfriend. I called his room in the morning to fix a time for a meeting and Ray asked for time to "take a shower." This was 11a.m. We finally met for a drink at 5p.m., went to a dinner and decided to meet the following morning. This continued for six days. On the seventh day I told Ray that Rachel must go back to teach at school and I had to return to my office in Tel Aviv.

"Okay," he said, "I am coming after you."

Sure enough, on the following Sunday he landed in Tel Aviv and we rented a room for him in the Accadia Hotel, on the beach in Herzliya.

The previous story continued the following week, except that during the daytime I could work in my office and not wait for Ray to "finish his shower."

On the following Thursday after we finished eating dinner I told Ray that on Saturday I must go to Ethiopia. As there was only one flight a week to Addis Ababa I had to take the flight on Saturday. We agreed, therefore, to meet on Friday morning in his hotel. We met. We spent one hour talking and we solved, satisfactorily, all the problems that needed a decision.

When I returned from Ethiopia a week later, I was told in my office that it took Ray another two days to pack his belongings and go back to Oklahoma. This story is God's truth and there is no exaggeration in it.

In spite of all the above, Ray Christian was the most successful Oil Well Service operator in the United States, where every hour of delay counts, as the well is not producing while the service is going on. How in the world he succeeded to be the most efficient operator in spite of his "punctuality" is the greatest mystery I have come across.

* * *

Barth Walker, Mayflower's chairman and lawyer, assisted me in Ghana during our negotiations with the government. He knew Accra from his service there during World War II as a US Air Force officer. It was useful to mention this fact whenever we met with the various government officials. There could not be two more contrasting individuals than Barth Walker and Ray Christian. Barth was a lovely man who ate just one meal a day, dinner. The strongest drink he had was milk. He never swore. He was never late. Yet, he and Ray were good friends and respected each other very much.

Years later, in 1977, my wife and I went with Barth and his wife, Lucille, for a trip through Austria and Switzerland. At first we were starving to death. No breakfast, no lunch. On the third day we decided: To hell with him. We are going to have both breakfast and lunch. We are not going to suffer the whole trip just to be nice. And that's what we did. They went for a walk and we had our regular meals. In the last few days Lucille defected to our camp. My wife would hand her several bread rolls which she ate with great relish without letting Barth see her committing this "sin."

* * *

As we had given Marvin Billet and Chris a participation in the Ghanaian venture they joined me a couple of times on my trips to Ghana. Marvin was great fun. His very well-developed sense of humor enlivened our long and hot days over there. He almost died on me there one day. Lunch had included some mayonnaise. In the afternoon we went to a beach club outside of Accra, where he started to feel really bad. His face turned green and gray. We managed to get back to the hotel, and with the help of the Israeli ambassador we located an Italian physician, Dr. Nochi. The doctor said: "You had mayonnaise for lunch, right? Two days ago the same thing happened to the German ambassador." The doctor prescribed three or four different medicines. When we asked him where would we find these medications in Accra, he turned around, put a white frock on himself, and said: "You is lucky. I am a pharmacist as well." Behind him was a long cupboard from which he pulled out his remedies.

From that day on Marvin named the chef at the Continental Hotel "Poison Pete," and a salad with mayonnaise "Nochi's special." Whenever we passed anybody having this delicacy, Marvin claimed that, at the bottom of the plate, below the salad, Nochi's visiting card was already lodged.

I played a practical joke there once. We returned to the hotel late one night. A breakfast order was hanging on the door of the room next to mine. I looked at it and saw that my neighbor had ordered two fried eggs. I added an X to all other egg dishes on the menu. There were probably ten different ones, boiled eggs, poached eggs, eggs with bacon, omelets, etc. The chef apparently did not have any sense of humor. The next day when we walked out for breakfast we saw the trolley being delivered to our neighbor's room. It was laden with every conceivable order of eggs. We did not wait for the outcome of the discussion.

Another amusing experience we had one night was when Marvin pointed out that this was Chris's birthday. Chris was resting in his room when we went to Accra's only nightclub, called La Rondo. Marvin started negotiating with a young lady asking her to knock on Chris's door. Her wrap-around dress, underneath which she was completely naked, would open up, and she would declare: "I am your birthday present." Whoever knows Chris, with his conservative, patrician bearing, can imagine this picture. We took her to the hotel and Marvin gave her final instructions. I was a little apprehensive that the Israeli ambassador might drive by and I would then have some explaining to do. Unfortunately, when we tried to enter the hotel with our "birthday present," the security guard at the hotel stopped us and refused to let her in. Chris's present was not delivered.

One day Marvin said: "You know I like this oil business. I am going to establish an oil company." I said that it was a good idea. "What am I going to call it? How about Accra (the capital of Ghana) spelled backwards?" he asked. This is how he created Aracca Petroleum, which was conceived in my room in Ghana. Marvin Billet told this story at one of his annual meetings in New York. My son, Kobi, happened to be there and Marvin said: "Gentlemen, if you do not believe this story, ask Zvi Alexander's son who is sitting in this room, and he will confirm it."

Major Zvi Alexander,
Israeli Army Signal Corps, 1952

Jewish soldiers in a unit of the Engineering Corps of the British Army in Egypt, engaged in smuggling arms for the Hagana. At right, Zvi Alexander; second from left, the poet Yehuda Amichai; in the center, Michael Koll-Nesher, Israeli Air Attache in Washington in the early 1950s

Officers of the Signal Corps with Chief of Staff General Yigael Yadin. In the middle, Chief Signal Corps officer Yaacov Yanai (Yan), to his right Itzhak Almog his deputy. Zvi Alexander third from the right, top row. Signal Corps H.Q., 1951

Meeting in New York with Colonel Yitzhak Almog, the former Head of Signal Corps, and his wife Mina, on their way to Japan to open an Israeli Defense Mission in Tokyo, 1954. The Almogs are seated at right, with Rachel Alexander in front & Zvi Alexander second from left

Mordechai Chen, the Managing
Director of Lapidoth, and
Zvi Alexander, his deputy,
Zahala, 1958

Visit of Levi Eshkol, the
Finance Minister, to the
Heletz oil field,
accompanied by Zvi
Alexander, 1960s

Eshkol during the same visit,
descending from the drilling rig floor.
At bottom, Zvi Alexander, 1960s

Finance Minister, later Prime Minister, Levi Eshkol, at Heletz oil field, accompanied by Zvi Alexander, 1960s

Visit by the Development Minister Yosef Almogi at Heletz. From left: Almogi, Mordechai Chen and Zvi Alexander, 1963

United Nations Seminar on Petroleum Exploration in New York, 1962. Sitting arrangement according to the alphabet: Last row, on the left: Zvi Alexander next to Dr. Said Amin, the Chief Geologist of Egypt. Seated at front, Zamihova, the representative of Madagascar who later became Minister of Petroleum of his country and issued a concession for oil exploration in Madagascar to I.N.O.C.

Group picture of all the participants in the UN Seminar. Zvi Alexander
is the first on the left of the second row, New York, 1962

On the left, Ray Christian, the President of Mayflower; his deputy, Bill Van Meter and Marvin Billet, the New York lawyer, alongside Christian's airplane. Oklahoma City, 1968

The signing ceremony of the granting of oil exploration licenses in Ghana. From left: Amos Ganor representing the Israeli embassy; Micki Deouell, our lawyer; Zvi Alexander; Sylvan Amegashi, the Oil Minister of Ghana. The last on the right is J.V.L. Phillips, the former Oil Minister, Accra Ghana, 31.12.1969

East Africa, late 1960s

The creation of PEDCO in Bermuda. Standing from left: Adi Ephrat, our
corporate lawyer; Zvi Alexander; an American attorney. Seated from
left: Dutch Lortcher, Chairman of Signal Oil & Gas; and Bill Miller,
Signal's V.P. of Legal Affairs, May 1972

Yitzhak Sirkin, Chairman of the Board of Directors of I.N.O.C., signs the oil concession in Madagascar. First from right, Micki Deouell, our lawyer, Tannarive, 1970

Anna Nicole Smith, the young wife of the billionaire J. Howard Marshall, our partner in Nigeria

Aracca went public at one dollar a share. It was a time of great enthusiasm on Wall Street for exotic oil stocks with licenses in dark Africa. A few weeks before the public issue, I called my friend and broker, Bill Hutchinson, in San Francisco to buy ten thousand shares at the issue price of one dollar a share.

I added: "If it goes up, buy me another ten thousand shares."

Bill Hutchinson answered: "Zvi, you cannot chase a stock. If you believe in it, order twenty thousand shares right away."

I agreed.

Aracca closed the first day at $1.75—up 75 percent. It continued its climb relentlesly and three months later, in January 1980, it reached a zenith of $10.75. I remember going to Hyde Park that morning and on the way picking up the *Herald Tribune* which had the quote for Aracca. It was $10.25, while the day before it had closed at $9.00. I waited impatiently for Bill Hutchinson to get to his office in San Francisco. I called Bill and asked about Aracca. The stock was at $10.75. I told Bill sell all twenty thousand shares when it reached $11.00, so that we would have made ten times our money—a thousand percent profit. These were famous last words. The stock never reached $11.00, and the next day it was back to $8.75.

There is a saying on Wall Street that bulls (those betting on stocks going up) and bears (those betting on stocks going down) both make money, but at different periods of the market. Pigs, the greedy ones, always lose. I am afraid that my involvement with Aracca stock puts me in the last category.

Luckily I started selling from $6.00 downwards and made money on the transaction. When Aracca shares went down to ten cents I bought a lot of them, believing that they couldn't go any lower. It's now one cent, which is only a nominal price. You can't sell it at any price.

* * *

Marvin Billet had an unusual way with beautiful women. His looks certainly did not justify it. He was of average height, a little pudgy, and not very handsome. Yet his success was phenomenal. I came to the conclusion that it was his great sense of humor that did it all. He turned boy-girl relations into a pleasant joke and it did the trick. One day we were having breakfast in Ghana and a good-looking blond woman walked in, an unusual sight at the time in Accra. We were still speculating who she might be, when we saw her again in the waiting room of the Israeli embassy, where we had a ten o'clock appointment. Marvin grasped the opportunity, stood up to introduce himself, and told a couple of jokes. Soon enough they were the best of friends. Her name was Marion Davis and she had come to buy diamonds in Ghana. She had an introduction to the Israeli ambassador and wanted to ask for his advice.

The next day Marvin took Marion to see the head of state, General Ankra. He had an introduction to General Ankra from a black American film maker, by the name of Bill Alexander. He therefore phoned the "Castle," the seat of Ghana's government, told Ankra's aide-de-camp about the letter of introduction and was invited to come that same afternoon. When our ambassador found out about it, he was furious. But I do not think that it did us any harm. On the contrary, it may even have helped us.

On another occasion, when we were travelling to Oklahoma City, Marvin had a serious problem with his leg, for which he was later hospitalized. He asked to sit at the back of the plane, so that he could stretch out his leg on several seats. During the flight I looked back to see how he was doing, but I did not find him. I walked back, and there he was, fully recovered, chattering and joking with a beautiful woman. Her name was Betty St. James, from St. Louis, Missouri. For years I kept her telephone number, in case I might be stuck in St. Louis. It never happened.

Once we were staying at the New York Hilton for a meeting with Ray Christian. The four of us, including Ray's assistant, went down late for dinner. There were two steps leading down to the dining room. We were waiting to be seated when we heard two blond girls, standing one step below us, asking the maitre d': "Do you have lobsters?" I suddenly heard Marvin saying: "Why have lobster? Have steaks." Embracing the two girls, he continued: "Waiter, a table for six." They were so dumbfounded that, before they knew it, we were all seated at a table of six. They were working for a brokerage house in Salt Lake City and had to deliver some securities to New York. Billet was soon the life of the party. Soon his friendship with one of our Salt Lake City guests came into full bloom. But before continuing with any more of Billet's escapades, let us go back to Ghana.

* * *

All company representatives arrived in Ghana after Christmas, 1968, for the signing ceremony scheduled for 9 a.m. on December 31. We were allocated four blocks in the easternmost part of Ghana, the Keta basin, two blocks onshore, and two blocks offshore, and we were very happy.

The representatives of Exxon and Gulf Oil Company did not receive a final confirmation from their head offices to sign the concession agreement. They tried desperately to get in touch with their bosses in the States, but due to telephone connection difficulties, they were unable to make contact before December 31. As we were all staying in the same hotel, each of us knew what the other was doing. Their problem was that the Ghanaian concession agreement had a so-called "Grandfather Clause" which stipulated that if the concessionaire offers better terms, to another African country, than those spelled out in the Ghanaian concession agreement, the Ghanaian government may request to renegotiate the current agreement. All the other companies were not greatly concerned with this clause, Exxon and Gulf were.

Gulf Oil was allocated the five choicest blocks of Ghana's offshore, the Tano Basin, on the border of Ivory Coast. When the signing ceremony was almost over, Minister Sylvan Amegashie stood up, looked at his watch and said: "Gentlemen, it is 10:26 Ghana time. I want to advise you that whoever did not sign for his allocated blocks will not be eligible in the future to apply for exploration in Ghana." He was the happiest of men being able to make this announcement to the two largest oil companies in the world, which had offended him by not signing.

After the ceremony I told our Ghanaian lawyer that I was most interested in the five blocks in the Tano Basin that Gulf Oil did not sign for. He promised to check with the competent authorities. He came back the next day and told us that there was a possibility to acquire four out of the five blocks. The fifth block must be given to Shell. In order to get the four additional blocks, we would need to make a certain contribution to the party. He mentioned a sum which for us was very substantial. I asked for a week's grace and promised to notify him by cable. I immediately flew home and told Dr. Dinstein about the Tano blocks and the requested "contribution." He said that he could not help me. If I could get the money elsewhere, fine and dandy. He, in any event, did not have any money for us. In spite of Dinstein's refusal, I cabled the Ghanaian lawyer that we were taking the four Tano blocks. I must have had a feeling that God would have mercy on me and that I would somehow, somewhere, obtain the necessary money.

Ray Christian arrived in Israel to conclude the operating agreement. In the meantime he bought out Billet's and Christensen's interests, and the two of us became the sole owners of the Ghana concession. I.N.O.C., through its Volta Petroleum subsidiary, which we had registered in Panama, now owned 54 percent and Mayflower owned 46 percent. We held a party at our home in Zahala for Ray and his group. Ray and I then went into another room and I told Ray that I needed a large sum of money. He did not ask me why and for what purpose. He said: "O.K. I will give you my 46 percent participation."

I replied: "No, Ray. I need it all."

"Do you really need it, Zvi?"

"I really do," I said. "You have it," he answered. Within a week, Ray transferred the whole sum. This was one of the few times that Ray's payment came in time. I told this story to Steve Christian, Ray's son, and assured him that I would not forget Ray's generosity to the day I die.

Chapter Five

Nigeria and the Story of Mark Rich

In late 1968 Dr. Dinstein was approached by J. Howard Marshall II regarding a joint venture in Nigeria. Nigeria was a large oil producer, producing more than a million barrels a day. It was a member of OPEC, the all-powerful cartel of oil producing countries. Nigerian oil was one of the best in the world. It was light, with almost no sulphur content. Nigeria was about to issue new licenses for the offshore areas, after a hiatus of several years caused by the war in Biafra. Almost all the oil in Nigeria came from the eastern region, the land of the Ibos, who had fought a war to create an independent country—Biafra. The Ibos lost the war and Nigeria remained one united country. Marshall suggested that Israel should try to obtain some of the new licenses; at the same time he would find large oil companies willing to take over after the granting of the licenses. These companies would operate and finance the venture, compensating us with free interest or a substantial royalty. Dr. Dinstein told Marshall to get in touch with me. It was a tremendous opportunity. If successful, it would promote us into a new league.

Marshall was one of the grand old men of the oil industry. His name was mentioned in a book describing the discovery of the mammoth East Texas oil field in the 1930s. He was then a young US Federal attorney who succeeded in overcoming the lawlessness and waste in the East Texas oil field. His picture appears in the book with the caption "The Champ." He later entered the oil industry and became president of Ashland Oil, then executive vice-president of Signal Oil and Gas, and subsequently president of Union of Texas Petroleum. At the same time, he served as executive vice-president of Allied Chemicals, which owned Union of Texas Petroleum. He retired at the age of sixty-five in 1968, the year he approached us.

When Marshall was with Signal Oil and Gas, he was the architect of secret commercial ties with the State of Israel, supplying Israel covertly from Signal's Iranian oil production. This story is described at length in chapter 17.

While I was busy concluding new agreements with Mayflower in Oklahoma City, Marshall flew in from Houston to meet me regarding the Nigerian venture. He was a very impressive individual, silver-haired and trim. He limped badly, as one of his legs was some six inches shorter than the other. Still, he played tennis almost every day. He was considered to be one of the wealthiest people in Texas. Among his holdings was a 10 percent ownership in Koch Industries, the largest privately owned oil refining company in the US. He was the only outside shareholder in Koch and a member of its Board of Directors. His stake in Koch, on its own, was worth close to half a billion dollars. Though he was twenty years older than I, he did not think twice about coming to meet me in Oklahoma City, as I could not go to Houston.

We had a very short and pleasant conversation and agreed on the terms of our joint venture. We would try to obtain the license in Nigeria, while he would find the "Big Boys" to explore and produce the oil. All expenses in the venture would be shared by us on a fifty-fifty basis.*

---

*   In the 1980s, an ugly quarrel broke out between Marshall and his only biological son, J. Howard Marshall III. The father accused his son of plotting to take over control of his shares in Koch Industries, dismissed him from all his posts, and wrote a will in which he left all his enormous assets to his stepson, Pierce Marshall.
Several years later, in his old age, Marshall met Anna Nicole Smith, who was his junior by sixty-five years. Anna Nicole Smith (whose real name was Vicky Hogan), was born in a small town in Texas, left school in ninth grade, became pregnant and gave birth to a son. She began to work as a nude dancer in a Houston nightclub. From there she went on to be featured as a "girl of the month" in *Playboy* magazine; she also became a model for revealing attire and an actress in soft porn movies. Apparently Marshall saw her for the first time in the Houston club and from then on kept in touch with her. In 1994 this acquaintance led to their marriage. At the time, he was eighty-nine years old, and Smith—twenty-four. The marriage lasted for fourteen months, until his death. During this short period, Nicole started quarreling with Pierce Marshall, the stepson, who called his father's young wife a "gold-digger," a term applied to young women who marry old men in the hope of benefiting from their wealth.
When Marshall passed away, his body was cremated, in accordance with the terms of his will. The stepson, Pierce, did not permit Smith to attend the funeral. She, on her part, turned to the court system and won a judicial victory: The judge ordered that Marshall's ashes be divided between the two—the stepson and the young wife;

Nigeria has the largest population in Africa with more than a hundred million people. I went to Nigeria at the end of one of my trips to Ghana. Thousands of people were milling around the airport in Lagos. They seemed very surly, and it was rare to see a smiling face. Everything looked dirty and crowded. What a difference from Ghana where almost everybody was smiling and pleasant.

Our ambassador in Nigeria, Ram Nirgad, took me to see the minister of transportation, Joseph S. Tarka. Nirgad and the minister were very good friends and addressed each other by their first names. Tarka was a very good-looking and well built young man. When I remarked, during the conversation, how fit he looked, he started doing push-ups, on the floor of his office to show me how he kept in shape. Tarka said that, to be assured of receiving an offshore license, he strongly recommended that our group should be different from other applicants. We would then stand a good chance of success. Otherwise the competition would be so fierce that he doubted he would be able to help us.

He suggested that we include local partners in our group. He wanted the local group to represent all three major tribes that formed the Nigerian nation: the Hausas in the north, who were Moslems; the Yorubas in the southwest and the Ibos in the southeast, who were both

and thus two funerals were conducted for Marshall, at each of which half the ashes were buried.

When the father's testament was opened, it became evident that he had disinherited his young wife and his only biological son, and had willed all his enormous property to his stepson, Pierce Marshall. Anna Nicole Smith immediately filed an appeal with the courts in California and Texas. She claimed that, right after the wedding, her husband had instructed his lawyers to change his will in her favor, but the lawyers had made contact with the stepson in order not to fulfill these instructions. After five years of very bitter legal wrangling, which provided the attorneys with millions of dollars in income, Smith won a clear victory: A judge in California ruled that, until a final verdict, she would receive half of the income from the assets said to be worth one billion and two hundred million dollars. Jurists were of the opinion that, in view of this ruling, chances were good for Smith ultimately to be rewarded half of the assets.

The marriage of Howard Marshall to Anna Nicole Smith and the struggle for the huge inheritance have in the last seven years become one of the spiciest items in the world media. Not only the yellow press but also *The New York Times* and the *Wall Street Journal* have devoted lengthy articles to the story.

All these events were still far off, of course, when I knew Marshall at the end of the 1960s. He was then in his prime and displayed a most expert knowledge and understanding of every phase of the oil industry.

mostly Christian. He wanted the local partners to pay their share, perhaps on a scale different from ours, but he did not want them to come in for free. Tarka emphasized that no other group was inviting local partners; therefore we stood a good chance of being considered very favorably. I said this was fine with me.

"You agree?" he asked.

"I certainly do," I replied.

"Then your first partner is waiting outside," he said.

He opened the door to his waiting room and in walked a very distinguished, portly gentleman, Chief Akin Olugbade, who was the former chief justice of the Federal Court of Nigeria. He was dressed in his native attire, with gold embroidery adorning his flowing white robe. We had a very pleasant chat and the next day we met and discussed some of the various financial commitments. We agreed with Minister Tarka that all payments would be made only after the granting of the license to our group. We decided to meet again three weeks hence, in Rome. By that time Chief Olugbade would, hopefully, be able to obtain the confidential maps of the areas to be released for licensing by the Nigerian government.

We also agreed immediately to form an offshore company, to be named Niger Oil Resources, which would be owned by Marshall and ourselves, with a 42.5 percent share of the capital each, and Chief Olugbade's group with 15 percent.

We arrived in Rome three weeks later. As Howard Marshall could not come, his partner in the Nigerian venture, John Crawford, took his place. We met with Tarka and went strolling with him on Via Veneto, the nicest shopping street in Rome. Micki Deouell, our lawyer, was with me. We turned into a side street at Tarka's request and climbed to a fourth floor tailor's shop. There were many bales of cloth on the shelves. Tarka kept pointing with his hand and the tailor brought down the indicated bale. I thought that Tarka wanted to inspect them at close range and then choose the ones he liked. Instead, when there were thirteen bales laid out on the table, Tarka said, "one of each," meaning thirteen suits, each to be made from one of the fabrics he had chosen. We then went to buy shoes made from different skins of lizards and crocodiles. It was some shopping spree!

We met again the next day to discuss various matters relating to our license applications. At the end of the meeting Tarka requested $10,000. We were flabbergasted. According to the agreement, no money had to be paid until the licenses were actually granted. Tarka insisted. He said: "You saw how many things I bought. I have to pay for them."

I protested: "Mr. Commissioner, nobody carries this kind of money with him."

"Travellers checks will do," he replied.

John Crawford, Marshall's partner, tried desperately to phone his bank in Houston, but it was already late, the bank was closed, and there was no way for the money to be in Rome the next day. We decided to inform Tarka accordingly. If it would kill the deal, so be it. We came to Tarka's hotel the next morning, fearing the worst. We told him that he could not suddenly spring such demands upon us. We would gladly have the $10,000 delivered to him in Washington, where he was going from Rome. "Washington will be fine," he agreed.

Howard Marshall reached an agreement with Ashland Oil, a very large independent oil company with headquarters in Ashland, Kentucky. Ashland was going to take over our deal in Nigeria when it would be ready for signing. In the meantime, we could use their name and submit all applications in the name of Ashland Oil. We needed this permission because neither Marshall nor we could even be considered as official applicants in Nigeria. The prerequisite for submitting an application was for the submitting company to have a strong balance sheet and ownership of a refinery. When Howard Marshall was president of Ashland Oil many years earlier, he had hired a young lawyer for the company, named Orin Atkins. Now Orin Atkins was the chairman of Ashland Oil.

Marshall and I came to see Orin Atkins in July, 1969 in his New York office. It was the first time that I met him. We reported on the developments in Nigeria and I expressed my confidence that we would be allocated two or three blocks out of a total of fifteen being opened by the Nigerians. It would be quite a coup. I had the map of the location of the new blocks that Chief Olugbade had given me in Rome. Atkins promised to send his people to see us right away.

The next morning at 8 a.m., two of Ashland's geologists walked into our room in the New York Hilton. They had taken the night flight in order to see us first thing in the morning. Micki Deouell, our concession attorney, was with me. We showed them the map of the areas to be released in Nigeria. Ashland's geologist pored over the map for a while, then said that all these blocks were located behind the "Sand Trend." They were therefore of very little interest. What he meant was that there would be no reservoir rock in that general area, and no sandstone layer which could contain oil.

This was one of the most stupid and ignorant geological statements I had heard during my life in the oil business. How a geologist, looking at a map of an unknown area, in which no wells had been drilled, could predict that there would be no sandstone reservoirs there, was beyond comprehension. Only after drilling a number of wells in any particular area can the presence or absence of sandstone reservoir rock be known. When that Nigerian area was eventually drilled, several years later, many sandstone reservoirs saturated with oil were discovered. Several very large oil fields were found in the area "behind the Sand Trend," in the areas that were originally allocated to us.

Orin Atkins could not go against his geologist's recommendation. Ashland Oil was a public company. A hired executive, no matter how high a position he holds, cannot go against the recommendations of his professional staff. If the area would prove to be barren, which is usually the case in oil exploration, the executive might lose his job. A week later we received a letter from Ashland Oil stating that, in view of their geologist's recommendation, they were withdrawing from the Nigerian venture. Several years later Ashland paid a multi-million dollar premium to get into a small section of the very same area. They probably fired their former geologist much earlier.

Howard Marshall replaced Ashland with two other companies: Penzoil, a large independent oil company from Houston, Texas, and Monsanto, the chemical giant. Our group received a free 13 percent carried interest, while Penzoil and Monsanto were going to fund the whole exploration effort.

Chief Olugbade, our partner, informed us that we were allocated two of the choice blocks in the Nigerian offshore. We were very happy. If oil would be discovered in these blocks, our 13 percent free ride would be worth many millions of dollars.

The representatives of Penzoil and Monsanto came to Nigeria to review the local situation. I accompanied them during their visit. We could not meet with Minister Tarka, as he was on an official visit in Poland. As both Jack Swink, the Penzoil representative, and Howard Karen, Monsanto's man, wanted to remain in Nigeria for another week to check drilling conditions, they suggested that I go to London to meet with Tarka, where he was stopping on his way back from Poland. We met at the St. James' Hotel. During the meeting, Tarka started to explain his whole political strategy to me. He had one major adversary, Chief Awolowo, the head of the Yorubas, the second largest tribe in Nigeria. I could not figure out why I was being introduced to all of Tarka's political maneuvering and scheming. I even said, "Chief (you could address him either as Mr. Commissioner or Chief), you should not tell me all these secrets." He said, "It is alright, you are a friend."

When I stood up to go he asked me to sit down again. He said that now, after the explanations relating to his campaign plans, I would understand his financial problems. In short, he needed a million dollars. This sum would be divided among several individuals and parties and would ensure the smooth passage of granting our concessions. The last $250,000 was going to buy a printing press, which he needed for his political campaign and for which he would give us detailed specifications. I almost fainted. It was all completely contrary to our original agreement. First, the sum was several times larger than the one we had originally agreed upon. Second, we had firmly agreed that all payments would be made only after the granting of the licenses. I walked out in a daze. My wife, who had come from Israel to meet me in London, was waiting in the hotel. When I walked in with an ashen face, she thought that I was having a heart attack.

I called Jack Swink, who was already back in Houston, and told him about the new development. He said that they were having an executive committee meeting that afternoon and he would call me back. When he returned my call at midnight, London time, he told me that Penzoil was pulling out of the deal. He also informed me that Monsanto would phone me in a couple of hours. At 2 a.m., Monsanto were on the line. They said that they might remain in the deal but would have to revise the agreement between us completely, obviously at much worse conditions for our group.

We made a new agreement with Monsanto whereby we were entitled to only 2 percent free interest instead of the former 13 percent. Monsanto took over the handling of the deal. The Nigerians also changed the terms of the new licenses, stipulating that the Nigerian National Oil Company was to receive a 51 percent interest in each license. When Occidental Petroleum, which was one of the four groups allocated blocks in this round (the others being a Japanese group, a German group and our group), agreed to the 51 percent participation, I told Howard Karen that these were going to be the new rules of the game. Once one company caved in and agreed to the new terms, everybody else would follow. Either Monsanto should agree to these terms or they should walk away. Instead they continued to argue. Howard Karen made thirty-nine trips to Nigeria. Finally, Howard Karen appeared in Lagos with a bank draft for three million dollars which was the signature bonus. It had to be paid to the government prior to the signing of the concession agreement. But it was already too late.

During this long period of two years, while Monsanto was arguing with the Nigerian government, Chief Akin Olugbade became more and more concerned that all his years of effort would be wasted if Monsanto did not sign the concession agreements. He finally gave Monsanto an ultimatum. If they did not sign within thirty days he would approach other companies. He made a deal with a French company for the two blocks. The French paid him one million dollars for the privilege. The joke was that the Nigerian government did not approve of the French company, as the French were then *persona non grata* in Nigeria. They had helped Biafra during the war. The chief kept the million dollars and sold the same license again to a Japanese group (not the group that had received blocks in the original allocation). He probably received another million dollars from the Japanese. Legally, part of this money belonged to us as it was paid to Niger Oil Resources, a company which we had formed and controlled. We wrote several demanding letters; we were also considering legal proceedings but did not pursue this avenue. Howard Marshall did not show too much enthusiasm for legal action, either, so we let it go. This was the end of our Nigerian venture.

* * *

In 1972, Ram Nirgad, the former Israeli ambassador in Nigeria, retired from the Foreign Service and was helping us in some other African matters, as a private individual. He met with Tarka at my request. He asked Tarka why he had made such outrageous demands from me in London. "Joseph, why did you do this to Zvi?" he asked. "Couldn't Zvi say no?" Tarka replied. Tarka was right. I should have known better. Even our meeting in Rome, when Tarka asked for $10,000, should have taught me that saying no to an African official is not a crime. It is an accepted procedure. But this was water under the bridge. You make a mistake, and you pay for it dearly.

* * *

During the following years I met Tarka on numerous occasions and we became good friends. It took me a while to understand that payments to politicians in the developing world are viewed completely differently than they are viewed in the West. A black American lady, a friend of my wife's, is married to a Ghanaian. They lived for many years in Nigeria. This lady was a lecturer in sociology at Chicago University and had a profound understanding and appreciation of African society. She once explained to my wife the different attitudes of Africans to the subject of payments to politicians. She said that in the West an able and enterprising man can make his mark and earn a lot of money in various endeavors, perhaps in industry, the professions or the arts. In Africa, on the other hand, and in many other developing countries, almost the only way to financial success is politics. The other opportunities of making money do not exist there as yet. In dealing with the developing countries, one should always remember this different set of circumstances and values, and the reasons behind it.

One also has to understand the local customs of social behavior in Africa. A man from the hinterland would not think twice before coming to the big city, and staying for many months with his well-off relative. This custom would also apply to a neighbor from his native village. The visitor would be housed and fed free without further questions. This is the extended family concept.

The waiting room of any important person in Africa is always filled with tens of needy applicants. They wait for an audience so that they can ask for a specific favor. This might take days, but eventually they would see the big man and make their plea. This custom requires a big house, many servants, and a lot of food. All this costs a lot of money and the money had to come from somewhere. The politician was also expected to distribute cash donations to his rural supporters. Tarka once told me about the sacks of money he used to take with him during his various trips in his constituency. This was the custom of the land and if you wanted to be in power you had to follow it.

* * *

I met with Tarka again in Rome in 1973, at the request of Signal Oil & Gas. Signal wanted to discuss with him the possibility of purchasing oil in Nigeria. Dov Ben Dror, my friend and former chairman, asked if it would be possible to introduce a representative of Phillips Brothers, the big commodity traders, to Tarka, for the same purpose. I agreed. A young man by the name of Mark Rich, arrived in Rome. This was the same Mark Rich who later became the most famous commodity trader in the world. It was not easy to make this introduction. Whenever Rich tried to enter Tarka's room, Tarka was busy with one of his lady companions. He was very active in this field. The servant guarding Tarka's door turned Rich back. Finally, on the third day I believe, we succeeded in making the necessary introduction.

Tarka travelled in style. He had a retinue of thirteen people travelling with him, including his high school teacher whom he wanted to honor. When we came to register at the Excelsior Hotel there was no vacancy. When we informed them that we belonged to Chief Tarka's party, all difficulties were solved on the spot.

I met with Tarka again, several months later, in London. Two of Signal Oil's top executives flew in from Houston for that meeting. The meeting was scheduled for 8:30 a.m. When we entered his suite at the Royal Lancaster Hotel, he asked us what we would like to drink. Before any of us had a chance to ask for coffee or tea, Tarka said: "Champagne is good for the kidneys." So, at half past eight in the morning, we started drinking Dom Perignion.

During those years, Tarka and his guests must have consumed a considerable quantity of Dom Perignion's output.

When, in 1977 and 1978, Nigeria was getting ready to return to civilian rule, Tarka was the architect of NPN, the National Party of Nigeria, which won the election. He probably would have been elected president of Nigeria, had he not been involved in a prolonged law suit involving a charge of gross corruption. Instead, he became a senator and the head of one of the most important committees of the Senate. In one of his campaign appearances in London he asked me to be present. He was later interviewed by the *Financial Times*. During the interview he turned to me repeatedly to confirm certain of his statements. He said, "This is my close friend and I trust him."

Tarka fell seriously ill in 1979 and became progressively more emaciated. We later suspected that he might have been one of the early victims of AIDS. He died in London that year. President Shagari of Nigeria sent a Hercules plane with more than fifty notables, headed by General Adebayo, the former chief of staff of the Nigerian Army, to bring his body back to Nigeria. A new university was built in his native State of Jos, the J.S. Tarka University, together with an eponymous town.

* * *

The story of Mark Rich, who was sent to Rome by Phillip Brothers to be introduced by me to Chief Tarka, has been well documented in various publications. He and his partner Pinky (Pincus) Green made a fabulous fortune in the late 1970s. Mark Rich was already a billionaire in 1980, when being a billionaire was still a very rare commodity. This was before the age of the Microsofts and the Internet billionaires.

Mark Rich and Pinky Green worked in Phillip Brothers, a company which was one of the largest commodity traders in the world. Phillip Brothers traded in various commodities but not in oil.

After the Six-Day-War, in 1967, a major oil pipeline was built to connect Eilat on the Red Sea with Ashkelon on the Mediterranean. The pipeline was half owned by the Israeli government and half owned by the Iranian government. The Iranian involvement was a state secret. The reason for the Iranian involvement was that Iran wanted an independent outlet to the Mediterranean, without being dependent on the Arab world.

Dov Ben Dror, my former chairman, became the President of the Eilat-Ashkelon pipeline. The throughput of the pipeline was many times larger than the consumption of Israel. This oil had to be sold on the world markets. The Arab boycott of Israel prevented the large oil companies and oil trading companies to buy this 'contaminated' (by Israel) oil. Ben Dror went to see Ludwig Jesselson, the head of Phillip Brothers, to convince him that Phillip Brothers should start trading in oil. They would make very good money buying the Israeli pipeline oil at an advantageous price. Once the oil left the Israeli port it is no more "contaminated." Oil does not have any identification marks.

Jesselson agreed and nominated both Mark Rich, who was representing Phillip Brothers in Spain, and Pinky Green, who was based in Switzerland, to start trading in oil. This is how Mark Rich and Pinky Green got involved in the oil business and eventually became the richest and biggest oil traders in the world.

At the beginning of 1974 I met Mark Rich again, in the lobby of our office building in London. We had an office apartment at 55 Park Lane, which we used as a mail drop address for oil companies that did not want to communicate with us in Israel. When I asked Mark what he was doing there, he answered that he resigned from Phillip Brothers the previous day, and is starting his own company.

In the early 1980s, Mark Rich and Pinky Green had problems with the US tax authorities. They moved to Zug, Switzerland, which became their headquarters. Mark operates, to this day, his commodity business from Zug.

Mark Rich donated hundreds of millions of dollars to various philanthropic causes in Israel. He supported education, university studies, hospitals, medical research and social activities, which would help to bring peace between Arab and Jew. He also assisted the Israeli Secret Service (Mossad) in rescuing Jews, whose lives were in danger, from Yemen and Ethiopia.

It is therefore understandable that Ehud Barak, Prime Minister of Israel at the time, and the head of the Mossad, Shabtai Shavit, were among the persons who recommended to President Clinton to pardon Mark Rich (together with Pinky Green) in the last days of his presidency. President Clinton pardoned both of the men, which permitted them to return to the United States without the fear of being arrested.

They would obviously have to first come to an agreement with the tax authorities in the USA. Experts believe that this long, drawn affair will eventually be solved by a compromise whereby they would have to pay taxes and fines of nine figures (hundreds of millions of dollars).

Although the presidential pardon met with strong censure by various groups, I, as one who has known Mark Rich for many years and know of his generosity towards Israeli causes, was very happy when the pardon was given.

Chapter Six

Uranium Exploration—The Patino Affair

At the beginning of April 1969, James (Jimmy) Ortiz-Patino visited Israel. He went to see the deputy minister of finance, Dr. Dinstein, regarding an air cargo company the Patino family owned. When Dinstein told him about I.N.O.C.'s mining concessions in Ethiopia, Patino asked to see me.

Jimmy Patino belonged to one of the wealthiest families in the world. His grandfather owned the fabulously rich tin mines in Bolivia, and until 1953 they were the tin kings of the world. In 1953 the mines were nationalized but, by that time, the Patinos had other mining properties around the world and a tremendous fortune in Switzerland. Jimmy and his brother George directed their world-wide operation from Geneva where they lived.

I met Jimmy Patino in Tel Aviv and it was love at first sight. He was very interested in our mining leases in Ethiopia, primarily in uranium. This was the time when nuclear power reactors were in vogue and there appeared to be an unlimited demand for future uranium supplies. We talked for several hours and agreed to stay in close contact. The day after he arrived back in Geneva he phoned me several times. At first we planned that our chief geologists would meet in Europe. But the Passover holidays were approaching and our head of exploration, whose father was a rabbi, could not travel. Jimmy then asked me to come instead and I agreed. He then said, "Let us finish this deal first and let the professional people sit down later." I could not have been happier. The deal was that the Patino Mining Corporation, then headquartered in Toronto, Canada, would spend one and a half million dollars on our mining leases, in return for a 50 percent interest in those licenses. We would remain 50 percent owners of the licenses without having to invest any additional money.

As the Passover school vacation had started, my wife, who was a teacher, joined me on my trip to London. The meeting was held at Patino's offices on the top floor of the IBM building in the City. I arrived five minutes early. When, on the dot at 12 o'clock, Erskine Carter, the president of Patino Mines, walked in, he apologized for being late. Carter was a tall, distinguished, and good-looking man in his late forties or early fifties. He asked to see our geological reports and feasibility studies. We had had these reports prepared by a world-renowned Denver company, Dames and Moore, and I handed them to him. He asked how much time he had to appraise the prospect and make a decision.

"How much time do you need?" I asked

"Would a week be all right?"

"Certainly," I agreed.

As the Easter holiday was starting and Carter was going to the south of France, we all agreed to stay in touch and meet in a week's time.

After the meeting Jimmy Patino went with me to my hotel—we were staying at the Westbury—to meet my wife. He apologized for Carter's behavior that would delay the conclusion of the deal. He said that we should humor Carter's whims. "It is only a week," he said. As I was going to Switzerland to meet with a United Nations geologist, an expert on Ethiopia, who lived in Lenzerheide near St. Moritz, Jimmy suggested that after my meetings we should come to him in Geneva, and wait there for Carter.

We rented a car and after a two day stay in Lenzerheide, we proceeded unhurriedly to Geneva. After our arrival, we were invited for drinks in Jimmy's home. He and his brother lived in a large compound on Lake Geneva, where each of them had a house with some additional buildings housing their offices. Jimmy's home was something out of a fairy tale. The fixtures in the bathroom were all made of solid gold. The waiter serving the drinks wore long white gloves, and there were many similar displays of great wealth.

The next morning Jimmy picked me up from our hotel to meet his brother George. On the way he told me that he was going to Paris to see Carter, as they had a very important meeting with the French government regarding nickel mines in New Caledonia. He also mentioned in passing that Carter would like to meet with me in Zurich, after their meeting in Paris.

I reacted instinctively: "Why Zurich? We are all here in Geneva, and Geneva is closer to Paris than Zurich. Why don't we all meet in Geneva?"

Jimmy said that I was right and he would so inform Carter. I did not give it another thought.

That evening we were invited for dinner by George Patino. I think that, except for George and ourselves, the guests were all titled. There were lords, counts, barons, etc., but we did not feel uncomfortable. During dinner, George, who was sitting near me, was called to the telephone. He returned after a long absence and told me, "That was Carter. I think I convinced him." He then relayed Carter's request that I should wait for his call at 11 a.m. the next day. I said, "Fine." I was not aware that something strange was going on.

The next morning my wife went out shopping while I waited for Carter's call. At 11 o'clock, on the dot, the phone rang and Carter was on the line. "Mr. Alexander," he said, "this is Erskine Carter, president of the Patino Mining Corporation." When I heard this formal introduction, I had a feeling of foreboding. "Mr. Alexander," he continued, "I am sorry to inform you that Patino Mining is not joining your projects in Ethiopia. I asked for a week and that week ends today. Thank you for your patience. Maybe we will do something together in the future. Goodbye."

When my wife entered the room, she found me in a state of shock. For the life of me I could not understand what had gone wrong. But somehow I started to get a feeling that it had something to do with the proposed meeting with Carter in Zurich which I had not agreed to. I made a reservation to return home on the first available flight. My wife insisted that I should call the Patinos first. I did so that evening. Jimmy was very apologetic and said that he was in the middle of drafting a letter to me. He and his brother were terribly sorry, but "when one has a good president, one must listen to his recommendations." I was invited to come and see them any time I was passing through Geneva. I met Jimmy Patino several times in the ensuing years, but the magic was gone.

The "Patino-Carter affair" continued to bother me for the next few months. In late July of that year I met with Bill Van Meter who was, back in 1967, Ray Christian's vice-president when we drilled in Israel. Bill flew from Dallas, where he was working for Exxon, to New York for a meeting with me, during which I offered him the position of I.N.O.C.'s vice-president. I believed, at the time, that we had solved our financial problems and I wanted Bill to work with us. I told Bill about our experience with the Patinos in Geneva. Bill suggested that I should go unannounced (in order not to be faced by a possible refusal) to Toronto and meet Carter. He felt that, until I had cleared up the circumstances of the Geneva debacle, it would continue to bother me. I flew to Toronto and called Carter. I remember the day very well, as it was a public holiday in the US due to the first landing on the moon by the astronauts.

Carter invited me to come and see him right away. I asked him what had happened back then in Geneva. I told him that I had not brought up the whole idea of cooperation with the Patinos. "How did it actually start?" he asked. I described to him the meeting with Jimmy in Tel Aviv and his subsequent phone calls from Geneva. Carter then said: "One day the phone rang in my home at 6:30 in the morning. Jimmy Patino was on the line. He had found the most important deal on earth. I should therefore drop everything and fly to London for a meeting. After I hung up, I said something to my wife. I am not going to tell you what I said but you can imagine it," he added. The ice was broken. We sat for several hours together and he explained his situation to me. He said that they were very busy. Not that one additional venture would break them but, in order to make such a decision, he needed a lot of time.

Carter added that he was concerned about the situation in the Middle East, and "Ethiopia is in the Middle East. How old is the Emperor?" he asked. Nobody knew exactly, but the consensus was that he was in his eighties. "What happens when he goes?" he asked. I replied that in Israel we were confident that the church and the military would form a coalition and take over after the emperor's demise, and that the same type of regime would continue. How wrong we all were! Carter also wanted to send a team of geologists to Ethiopia to check the situation on the ground. For all of this he needed several months, and not just one week. He had planned to explain all of this to me in Zurich. When this meeting did not materialize, he decided to terminate the project. With great sincerity, he invited me to apprise him of any new ideas in the field of mining we might come up with in the future. We parted good friends.

We allowed the mining leases in Ethiopia to lapse, since we did not find other partners for them.

I often thought that this "Patino-Carter Affair" should be studied in business school. I was planning to present it to the Columbia University Graduate School of Business but never got around to doing so. One of the most important lessons to be learned from the story is that you should always try and negotiate with the correct level of management on the other side. Even when you agree on a deal with the owners of an enterprise, their official president, if he is worth his salt, will try to kill it. Do not negotiate at too high a level, and obviously not at one that is too low. If you try to conclude a deal with too high a level, the correct level will do everything in its power to destroy it. Always try to find the level responsible for the particular subject matter you want to discuss.

I later thought to myself that, if Dr. Dinstein had wanted me to make a deal with somebody and would then insist on participating in the meetings and in the decision-making, I would have done everything possible not to allow the deal to be completed. It is just plain human nature. I have since tried to follow my rule of the "Correct Level." Whenever this was not possible, or I failed to recognize the correct level, that deal did not go through.

Chapter Seven

Wall Street—Jewish Bankers, Arab Money

It was the beginning of 1969. We already had large oil concessions in Ghana and Ethiopia. We had three mining leases in Ethiopia and were actively looking at other places. We needed money, and lots of it. The only logical source for this type of risk money was Wall Street. From that time on, most of my attention was directed at satisfying this urgent need for funds.

As Equity Funding, our partners in the Ethiopian venture, was one of the brightest stars on Wall Street, I explained our problems to Herbert Glaser, one of Equity's directors, and asked for his help and advice. He suggested that I should see Robert Nagler in Geneva who was a partner in Oppenheimer's, an investment banking firm in New York. I met with him in January 1969 in Geneva. He liked my story and wrote a letter to Leon Levy in New York, who was a senior partner in the firm, recommending that the house of Oppenheimer undertake the financing of I.N.O.C. We wrote to Levy but did not hear from him for a long time. Then we found out that he had been skiing in Switzerland for a month. As a result, we had missed out on what turned out to be a very crucial period of time. Wall Street was flourishing, money was most plentiful. It was a perfect opportunity for raising money for speculative issues like oil exploration in exotic countries. All this we did not know and, even if we had been aware of it, we would not have known how to go about it.

There was at that time only one Israeli company that did raise money on Wall Street. This was Maritime Fruit Carriers. They were a privately owned, powerful company. They had modern, ocean-going, refrigerated ships, while we were a penniless government-owned company with unexplored oil concessions.

The Israeli government ownership of our company was another serious obstacle. Wall Street did not like government-owned entities which were usually not very efficient and not renowned for profitable operations. Therefore we had many strikes against us, in addition to the Arab boycott, which was most effective in the oil sector.

In April of that year, a prominent New York lawyer by the name of Hans Frank visited Israel. Frank was an acquaintance of Hanan Yavor, the former Israeli ambassador in Ghana and Nigeria, a director of I.N.O.C., and my dear friend. Hans Frank was a senior partner in one of the largest and most influential law firms in the United States—Fried, Frank, Harris and Shriver (Shriver was the brother-in-law of President John F. Kennedy). In spite of Frank's thirty or more years in New York, his accent and demeanor immediately disclosed his German background. He heard my story and told me that his closest friend was a banker by the name of Mark Millard. Millard was a senior partner in Loeb Rhoades, a very prestigious investment banking firm in New York and a specialist in oil venture financing. I would not be able to find a better introduction. I prepared all the necessary documents, describing our various holdings, and proceeded to New York.

Unfortunately Mark Millard was away in Argentina for ten days. Instead, I met with his two assistants, Sydney Knafel and Steven Patchek. We all got together with Hans Frank for dinner at the Plaza Hotel and again the next day in the Loeb Rhoades offices. I had the distinct feeling that Millard's assistants did not understand what we were doing. Hans Frank's fertile brain was structuring various modes of financing based on "Tax Money," whereas I was much more interested in equity capital. Only equity capital would give us a solid base for future growth.

The concept of "Tax Money" was very important in the American economy. By law, many different investments were allowed to be charged against current income, thus reducing the tax liability of companies and wealthy individuals. In effect, therefore, the American government covered 50 percent of such investment when the investor was in a 50 percent tax bracket. Direct investment in a specific oil exploration venture belonged to the category of "Tax Shelters," whereas investments in shares of an exploration company were not tax deductible. It was therefore quite difficult to find equity investors for an oil exploration company.

But this was exactly what we needed to start building a viable company. Tax money could only be invested in specific exploration projects. It could not be used for general corporate purposes. To build a company, we needed equity capital. Once we had a company, we would be able to approach the financial markets to raise additional capital for future developments.

Several days after the meeting with Loeb Rhoades, Kenneth (Ken) Bialkin came to see me. I had first met him in 1966 when I came to New York for the Hirshhorn-Livingstone consortium meeting. Rami Taiber, a director in Lapidoth, recommended that I engage Bialkin's services to participate in that consortium meeting. Rami had met Bialkin in the early sixties, when Rami was stationed in New York, representing the Israeli government. He was very much impressed with Bialkin's abilities. Bialkin was then a fairly young attorney with Wilkie Farr, one of the most prestigious law firms in New York. I saw him from time to time when passing through New York and we became good friends. But as most of our business during the period of 1967–1968 was in Oklahoma, we did not use his services. When I saw him again, on this trip in 1969, Ken Bialkin was already one of the senior partners in his firm. He later became one of the leaders of the Jewish community in America, the President of the Conference of Presidents of Major Jewish Organizations, a leader of B'nai B'rith, and one of the great supporters of the State of Israel. Ken became I.N.O.C.'s lawyer during this first money-raising trip of mine and remained our legal representative, friend, and supporter for many years.

He was a trim young man when I first met him. Since then, in thirty-seven years, he has not gained an ounce of weight. He is the only one of my acquaintances with this outstanding record. He is quiet, very polite, and a hell of a nice guy—the complete antithesis of the loud and brusque prototype of a New York lawyer.

I told Bialkin about my unsatisfactory meeting with Loeb Rhoades. Bialkin said that most of the investment banking firms in New York were his firm's clients. He did not think that it would be very difficult to arrange a meeting with any one of them. I mentioned with trepidation the name of Lehman brothers an investment house in existence, for the last hundred and fifty years. Lehman brothers, in addition of being the strongest and most prestigious investment house on Wall Street, were also very much involved in the oil business. Bialkin promised to try. The next day he arranged a meeting for me with Don Russell of the Lehman Brothers oil division.

Don Russell was a silver-haired gentleman in his late fifties, courteous and soft-spoken. He was a founder and part owner of a small Canadian oil company called Sunlite Oil and was very knowledgeable about the oil business. As Lehman Brothers were financing many of the international oil companies, Don Russell was also very familiar with exploration activities in Africa and the rest of the world.

While I was having my first conversation with Don Russell, the senior partner in charge of Lehman Brothers' oil division, Edwin Kennedy, a legendary banker, walked in. In a book published in the US in the early 1960s called *The Money Magicians*, a whole chapter is devoted to Edwin Kennedy's exploits.

Kennedy said: "Mr. Alexander, I have only fifteen minutes. Could you tell your story in fifteen minutes?"

"I will try, Mr. Kennedy," I replied and I finished in thirteen minutes flat. I tried to include in my account what we were doing, our mode of operation, where we were headed, what were our goals, and why I was looking for money.

When I finished, Kennedy asked: "Where did you go to school, Mr. Alexander?"

"I went to the Columbia Business School, Mr. Kennedy, but actually, my school was the Israeli Army."

"Oh, you are making me change my views about the army," he concluded and left shortly thereafter.

The army was not a very popular subject in the US in 1969, in the middle of the Vietnam war. If a man like Edwin Kennedy was making such a statement, then I must have made a good enough impression on him. There might be some hope that Lehman Brothers would be interested in helping us.

Don Russell took me to lunch and we continued discussing our financial needs. He asked me how much money I was looking for. I told him that we were looking for twenty million dollars (which would buy half of the company). He then asked what the appraised value of I.N.O.C.'s foreign holdings was. I replied that it was in the neighborhood of ten million dollars. To his question as to how I arrived at this figure, I answered that Ashland Oil had paid a million dollars for a 10 percent interest in an offshore concession in Togo, which is just due east of our four blocks in the Keta basin in Ghana. As we owned 54 percent of a much larger concession in Ghana, and as there was no reason to consider our area any less valuable than Ashland Oil's area in Togo, using the price Ashland paid as a yardstick, the value of our Ghanaian holdings should be at least seven million dollars. Russell agreed. He asked whether I used the same criteria to appraise our other holdings. I replied in the affirmative. Russell then inquired as to what figure I reached. I said it was ten million dollars.

"Why are you then asking your investors for twenty million for a 50 percent interest?"

"Mr. Russell, if I have achieved this asset accumulation in less than eighteen months, without any money, don't you think that we are entitled to ten million dollars of goodwill?"

"I think you are right," he replied. In this way the value of I.N.O.C.'s holdings was established as twenty million dollars, and this figure formed a basis for future negotiations with various investors.

After we finished eating, Don Russell told me that they would need several days to consider our case. I had to realize, he said, that they were very heavily involved in the Arab world. They could not do anything to jeopardize their standing there, nor embarrass any of the international oil companies operating in the Middle East or North Africa, who were their clients. If they would agree to finance us, it would have to be done quietly, without too much publicity. He added that there were many ways of doing it confidentially. After all, twenty million dollars was not a large sum for Lehman Brothers.

Before I left, Don Russell asked me for the names of two people who had done business with me. I gave him the names of Ray Christian, my partner in Oklahoma City, and John Christensen, the geologist from Rome, who happened to be in New York, and whom Don Russell knew. I found out later that before I arrived back at my hotel, Russell had already checked me out with both individuals.

It is quite incredible that the appraisal of our assets at a twenty million figure was reached casually at a lunch table, and that this sum later formed the basis for Lehman Brothers' financing effort. A year later, when I was trying to obtain a half million dollar investment from an insurance company in Chicago, at a time when our assets were at least twice as large, and when a large independent oil company had already invested millions of dollars in our company, proving its full value, the insurance company spent two solid months appraising our assets, asked for hundreds of documents, and finally requested nine separate personal references. Each of these referees was actually contacted by this insurance company—and in the end they did not invest.

But, in 1969, the market was booming. Various mutual and hedge funds were raising hundreds of millions of dollars from investors and were looking for a home for the moneys they had raised. Foreign oil exploration was one of the most exciting games in town. There was a lot of enthusiasm for the newly-independent countries with exciting mineral possibilities. The sky was the limit. But 1970 was totally different from 1969. 1970 was the year of a stock market collapse.

Don Russell called me two days later, informing me that he had arranged a meeting with a prospective investor, and asked whether I would be free at 10 a.m. for such a meeting. This was the usual low-key approach used by Don Russell and other members of the firm. Would I have time? Really!

We went to see Mr. Becker who was the managing partner of a hedge fund called Fleshner-Becker Fund. A hedge fund is a fund that simultaneously buys stocks believed to be going up, and sells other stocks short.* Thus if they guessed correctly, their profits would be double of those of regular funds. They made money when the stocks they believed in went up and also profited on the shorted stocks, on their way down.

I remember Becker's office very well as it was the only round office I had ever seen. It was a complete circle. Becker was a stocky man in his early forties, and he received us very graciously. From the beginning, I seemed to be completely unneeded at this meeting. The whole conversation took place between Don Russell and Becker. I was asked one question only throughout the hour and a half the meeting lasted. Becker inquired why the African countries were voting against Israel at the United Nations, when we claimed to have such wonderful relations with them. I explained that many Arab and Muslim nations belonged to the OAU (the Organization of African States), and that many African countries needed the Arab vote to further some of their own projects. In reciprocation, the Africans voted with the Arabs against Israel. But this voting did not reflect reality. Our relations with Africa had never been better. Israel's assistance in agriculture, health and military training was flourishing. Becker accepted this explanation which, I believe, was my only contribution to the conversation.

Russell and Becker started discussing the possibility of Becker's investing four million dollars for a 10 percent interest in "Oil and Minerals," our offshore operating company. Our corporate plan was, from the beginning, to form an offshore company which would be the operating company for our world-wide effort outside Israel, and in which

---

* "Selling short" is a system whereby an investor borrows the stock that he wants to sell short, and sells it. He does it in anticipation that the stock would go down soon. When the stock actually goes down, he re-purchases the stock at a lower price and returns the stock he borrowed. As he originally sold the stock at a higher price, he pockets the difference.

I.N.O.C. would be only a stockholder. This would be the vehicle where foreign investors would make their investment. It was perfectly clear to us that, for many reasons, nobody would be willing to invest in an Israeli-registered company. One of the main obstacles was the Israeli laws that restricted free transfer of currency. Israeli taxation was another obstacle. Not only were taxes punitively high, but there was also no reason to pay taxes in Israel on profits made in Ghana or in any other foreign country. The political risk was another factor. An additional important consideration was the advantage in operating abroad under a neutral flag, in order to diminish the pressure of the Arab boycott. For all these reasons we planned to form "Oil and Minerals" and register it in Bermuda, as soon as we would find the first investor.

In the Lehman-Becker case, I.N.O.C. was going to transfer its foreign assets to Oil and Minerals for twenty million dollars' worth of stock and hold a 50 percent interest, while the new investors, including Becker, would invest twenty million dollars in cash for their 50 percent interest. Becker made one condition. This was that Lehman Brothers would agree to take the new company public, and list it on one of the stock exchanges, within the next eighteen months. Becker's Fund needed such listing on a stock exchange for a current valuation of their portfolio of stocks. Don Russell said that he would have to check it with Lehman's syndication department, which was in charge of public issues.

Going down in the elevator, after the meeting with Becker, I had a feeling of unreality. I asked Don Russell for reassurance. He said that they (Lehman Brothers) had made nineteen million dollars for Fleshner-Becker in the previous year. As a result, Becker was well advised to listen to them. We went back to Lehman's offices, which occupied a whole building near Wall Street. Don Russell called in the syndication department for consultation. He described the company and Becker's request for public listing. Their reply was that, subject to the stock market remaining unchanged, Lehman Brothers would see to it that we went public within the prescribed time. Although Lehman themselves would not do it, because our company was too small and because of the Arab boycott, there were enough satellite, smaller investment houses that would tackle the public issue at Lehman's request.

We then discussed Lehman's remuneration for this transaction. We agreed on 6 percent, to be paid in the shares of the new company. This was a very complimentary gesture, indicating Lehman's confidence in our success. It was also very important for us to have Lehman Brothers as our stockholders. We could not ask for a more prestigious and powerful name. Don Russell called Becker, informed him of the syndication department's reply, and arranged a meeting for the following day.

This meeting with Becker was very similar to the previous one. I was not asked one single question. At the end of the meeting, Don Russell said to Becker: "We have you for four million dollars." Becker replied: "You have a deal."

We shook hands and walked out. I asked Don Russell what kind of a commitment "you have a deal" was. He replied: "Mr. Alexander, when somebody on the Street (meaning Wall Street) tells us, 'Lehman Brothers', 'you have a deal,' you do have a deal! Spend it wisely," he added.

Ten days later, when I was already in Israel, I was informed by telephone by one of Ken Bialkin's associates that he was holding in his hands a contract, signed by both Lehman Brothers and Fleshner-Becker. It called for the investment of four million dollars to be paid in four equal monthly instalments, starting on August 1, 1969.

The funny thing was that I did not feel very different. Here I was, with hardly enough money to pay for my airline ticket and hotel bill, and wondering whether we would be able to meet the following month's I.N.O.C. payroll, while at the same time I was discussing the creation of a forty million dollar company of which we would own 50 percent. What had happened was a miracle. The legendary Lehman Brothers had agreed to be our bankers. We had overcome the Arab boycott, the greatest barrier to our progress. We had solved the financial worries which had accompanied us from day one. We were going to have ample funds to explore anywhere in the world. I was in seventh heaven, riding on a cloud. But apparently the nature of the human being is such that it takes a very short time to get used to improved circumstances. This is also what happened to me.

I met with Don Russell again the next day, when he suggested that we raise only eight million dollars in this first effort. This would give us enough money to operate comfortably for a year or two. As the company would develop, in the interim, we would be able to charge a higher price per share for the remaining twelve million dollars. Obviously, I agreed.

The other prospective investor introduced by Lehman was a former Israeli who had made a fortune in the States importing electronic equipment from Japan. He was introduced to Don Russell by an Orthodox Jewish broker, who was acting as Russell's bird dog. I did not like this investor and did not feel comfortable with him. I had great doubts that he would eventually invest. Oil exploration was not an Israeli type of investment, nor was it compatible with electronics sales. I expressed my reservations to Don Russell. He admonished me by saying: "Mr. Alexander, raising millions of dollars is not an easy task. Would you please leave it to us."

There were no clouds on the horizon. Don Russell made preparations to go to Paris, where he was to help distribute shares of "Natomas," a young oil exploration company which had exciting prospects in Indonesia, and which was sponsored by Lehman Brothers. I had to travel to Ethiopia. We therefore agreed to meet in a couple of weeks, in June, to conclude an agreement with a second investor for the remaining four million.

I had two other important meetings in New York during that trip. One was with Frank Creary, a principal in a company called Whiteshield. Creary was nicknamed "the man with the golden hand," as, in 1968, he had personally raised sixty-six million dollars from wealthy individuals for oil exploration. This was a tremendous achievement. Creary was excited about our prospects in Ghana and wanted to buy 25 percent of our concession. We agreed on a price of five million dollars. Next day he called me to say that my deal was "all over town." I then found out that Ray Christian, my partner in Ghana, had concluded a deal to sell 50 percent of his 46 percent to a group of young bankers in New York for $460,000. He needed the money for the seismic survey in Ghana. As he had other business to take care of, he did not try to find the best buyer and sold his interest for a very low figure. Our bad luck was that Ray's sale was publicized at the very time that we made the agreement with Frank Creary. Creary terminated his negotiations with us.

The other important meeting was with Mark Millard, the senior partner in Loeb Rhoades, who was the banker I originally came to New York to meet with. We finally met upon his return from Argentina. He was a most impressive individual, tall and wide-shouldered, with black, bushy eyebrows and a booming voice which, at the same time, was very melodious. I told him about my fruitless meeting with his assistants and Hans Frank's insistence on "Tax Money" deals which could solve only part of our problems. He laughed and agreed with my remarks. I told him about our agreement with Lehman Brothers. He was tremendously impressed. He repeated twice: "Only Lehman could have done it. You're on your way, Zvi." He offered to help us with our mining leases in

Ethiopia and to arrange a meeting with a large German company called Metallgesellschaft, for which I was very thankful. We parted good friends.

When I arrived home, the employees of I.N.O.C. were very happy. They had been constantly worrying about the stability of their employment and the long-term survival of the company. They received the reassurance that they so badly needed and acquired a feeling of achievement and security. Various government officials congratulated me. Still, I do not think that any of them realized the importance of the breakthrough we had made, except for Dov Ben Dror, my former chairman. He told the finance minister that, in his view, our agreement with Lehman Brothers was the greatest economic achievement since the State of Israel had come into being.

Several days later I left for Ethiopia, where I had to stay for a week. Upon my return, I contacted Don Russell who, in the meantime, had returned from his trip to Paris. The mood on Wall Street had started to turn sour, but Russell was confident that the second investor would be found shortly. The Israeli investor, in whom I had great doubts from the beginning, disappeared.

I went back to New York during the 4th of July weekend. Adi Ephrat, our corporate attorney, came along. We stopped in Frankfurt and met with Prince von Wittgenstein, the head of Metallgesellschaft, to whom we were introduced by Mark Millard. We presented our mineral licenses in Ethiopia. We had a pleasant meeting, but nothing came out of it. We arrived in New York on July 5, 1969, and met with Don Russell the next day. He told us that Becker was extremely upset by the fact that no additional investors had been found, as yet. The fact that the market had taken a turn for the worse and Becker's fund must have suffered some reverses did not help. Russell asked me whether I could have any of my friends go and see Becker and assure him of the soundness of the investment. John Christensen, the geologist from Rome, happened to be in New York and went to see Becker. Chris had a very distinguished appearance and, although he used understatements, he was very persuasive and convincing. He tried to explain to Becker the great potential of our company, but apparently without much success.

Several days later we were invited to go and see Becker. We all went together with Don Russell. Howard Marshall, our partner in the Nigerian venture, agreed to join us also, in order to add weight to our delegation. Becker, his partner Fleshner, and another director of his fund were present. Becker opened and said, "Mr. Russell, are we the only investor in town? We know for a fact that you have talked to many people on the Street (Wall Street). Why did you not find other investors?" He added that Lehman had promised that there would be an oil investor in the deal, as well. "Where is he?" he asked.

Russell explained that he had gone to France while I had gone to Ethiopia and therefore it was going to take more time. As to the oil investor idea, it would be very difficult to find an oil company investing in shares of another oil company. Here Howard Marshall intervened to explain that because of the structure of the American tax laws, it was most unusual to find one oil company investing in another company's shares. One would find only a handful of such cross-investments among the thousand or more oil companies operating in America. Marshall said many good things about our concessions and the way we went about acquiring them. Marshall's eminent standing in the business community in the US (he was reported to be the wealthiest man in the State of Texas) added considerable credence to his statements. Still, all of this was not of great help. As Zevi Dinstein used to say, "You have either success or explanations." We left the meeting with Don Russell promising to find the other investor or investors as soon as possible.

I have often compared Becker's feelings to those of a well-to-do lady who is contacted by her favorite shop before the annual sale commences. She is offered a beautiful dress at a reduced price and she is very happy. Several weeks later, when she passes by the shop, she sees the very same dress hanging in the window, unsold. She will never wear the dress again, no matter how much she liked it before. Becker believed that Lehman Brothers brought him the very best deal in town. With every passing day, when other investors did not materialize, he started losing confidence. He began having doubts as to whether he had made the right decision. Furthermore, the problem of the lacking oil investor was a serious one. We were faced with this problem again and again. If what we had was so exciting, why did not another oil investor invest in our company? Howard Marshall's explanation to Becker, why such an investment is not the accepted practice in America, was difficult to convey to the non-oil investors.

We had other contacts with potential investors during that period. One was quite amusing. Adi Ephrat and I went to see Harvey Krueger, one of the senior partners in Kuhn Loeb, another prestigious investment house in New York. Krueger showed me a prospectus for debentures of Bank

Leumi, the largest bank in Israel, in which Kuhn Loeb participated or was even the lead underwriter. Krueger said: "You do not have to go to Lehman. We are not afraid of the Arab boycott. You should have come to us." I said, "Fine, here I am." Several weeks later I was asked to meet with Harvey Krueger again. When I came, I was taken to see another senior partner. He said: "We have found a solution for you, but you would have to give up 60 percent equity to our investor." I asked who that investor was. The senior partner replied that it was Burmah Oil. I could not believe my ears. I said, "Burmah Oil owns more than 20 percent of British Petroleum, which derives most of its income from the Arab world. Burmah Oil would not come anywhere near us for fear of the Arab boycott." "Mr. Alexander, they are our clients, they always listen to us," said the senior partner. I had no further dealings with Kuhn Loeb or its partners in the last twenty years. But I met Harvey Krueger again in 1998 during the 45th annual America-Israel Chamber of Commerce testimonial dinner. Our son Kobi was the guest of honor and Harvey Krueger was the master of ceremonies. By that time he was vice-chairman of Lehman Brothers, who had bought his old firm Kuhn Loeb. During the presentation, Harvey Krueger mentioned that thirty years previously he had assisted Kobi's father, who was sitting at the table, to raise money on Wall Street.

* * *

Adi Ephrat, who was our corporate attorney, stayed with me through all these long summer days in New York. We were always waiting for the crucial telephone call. I do not think we went out more than a handful of times. We were waiting for the next telephone call which, we hoped, was going to solve one of the pressing problems of the moment. Adi was not only our attorney and the best lawyer I have ever come across, but also a friend, advisor and confidant. My long stay in New York would have been much more trying had he not been with me, counselling me and soothing my frayed nerves all the time.

Lehman Brothers continued to look for an additional investor. It was becoming more and more difficult from day to day. The stock market went from bad to worse. Mid-July turned into a hot and humid August in New York. Very many people were on vacation in far away places. Finally, in late August, I was informed verbally by Don Russell that, regretfully, Lehman Brothers were pulling out of the deal. Their efforts to raise money for us were now widely known. They had originally planned to do everything in a very low key manner, since they could seriously suffer from unnecessary publicity. Their ties with many oil companies working in the Middle East and North Africa might be put in jeopardy. They wanted out.

We did not know what to do. We had legal documents binding Lehman Brothers and Fleshner Becker to us. They were not conditional on other investors joining the deal. Fleshner Becker had to put up the first million dollars on August 1. We spent many hours with Ken Bialkin and two of his colleagues weighing various options. We wrote Becker a letter demanding payment. Becker replied that Lehman promised him an oil investor, and as this did not materialize he found himself free from obligation. We considered legal action. But how can you sue somebody who does not want to invest money in you anymore? A law suit could take years. What do you do in the meantime? How would Wall Street view you when they found out that you were suing your investors? Would not such litigation scare off potential future investors? With these depressing thoughts, Adi and I returned home in late August.

We were greeted with disbelief and even scorn. "We told you that you were dreaming," many people in government said. "We do not believe that the Arab boycott has much to do with it," they added. "All the ideas about raising money on Wall Street were hallucinations and daydreams of Zvi Alexander," was the statement going around in the Ministry of Finance. Then John King's big attack on us started. That story and how we survived it will be told in the next chapter.

Ken Bialkin continued his contacts with Lehman Brothers on our behalf. Some time in September he received a letter from them, signed by Edwin Kennedy, citing the Arab boycott as the reason for their withdrawal from our financing. It was a silly and naive move by Lehman Brothers to put such a statement in writing. Giving in to the Arab boycott was illegal in the United States. The damage that Lehman Brothers could have suffered if such a statement were to be spread across the front page of *The New York Times* would have been very considerable.

With this letter in hand, the government in Israel could no longer ignore my problems. This letter called for government action. The economic minister of Israel in the United States was instructed to make an official complaint to Lehman Brothers, to warn them of the illegality of their attitude and demand reinstatement of their involvement.

Lehman Brothers realized the mistake they had made by putting their fear of the Arab boycott in writing. They decided to try and revive our financing. At that time I was in contact with a potential investor by the name of Theo Ben Nahum. I had first met Theo when I came to New York in late April that year, to find an investment house. He was very polite, but he did not pay much attention to me. When he found out later about the Lehman involvement, it was a different story altogether. If Lehman Brothers found our company of sufficient interest for their involvement, then our company merited attention. The name of Lehman Brothers was magic. Their stamp of approval carried immense weight.

Theo Ben Nahum was an ex-Israeli who had emigrated to the United States before World War II. He was a close relative of Zevi Dinstein's wife, and Dinstein had suggested originally that I should see him in New York. Theo did well in the States. He owned an entire floor of a brownstone off Fifth Avenue, adorned with many expressionist paintings. He was connected to the Allen brothers who had a relatively small but extremely successful investment house which carried their name. Together with Allen and Company, Theo formed a small oil company called Planet Oil. Planet Oil had almost no assets except an oil concession in a godforsaken country called Mauritania. But they also had a very important and lucrative concept on which they were working diligently. It was to increase the quota of Iranian oil to be imported to the US. As Iranian oil was selling at $1.50 a barrel when oil produced in the United States was selling at $3.50, a new quota of imported oil would have been a license to print money, a lot of money.

When I met Theo again in October 1969, Planet had spent most of the ten million dollars they had raised initially. They had about one and a half million dollars left in the treasury. The plan we developed with Lehman Brothers was that Planet Oil would invest one million dollars in Oil and Minerals. This investment would revive Planet Oil's fortunes by showing its investors the exciting concessions we had. Planet's investment would satisfy Fleshner-Becker as to the presence of an oil investor in the company, who would then invest $2,200,000. Lehman Brothers would also invest some money from its own funds for a total capitalization of eight million. The market had changed dramatically. Good-bye to the ten million dollars goodwill; Good-bye, forty million dollars capitalization.

We all met at the Lehman offices, and the new financing plan with Planet Oil and Becker was agreed upon. Theo Ben Nahum sought Charlie Allen's approval for the deal. Allen suggested that Theo, instead of spending nearly all of their remaining cash resources, should rather sell a block of shares of Planet Oil. He added that he did not mind at what price. Let Theo find an investment or pension fund and place these additional shares privately. As long as he, Theo, could place these shares and raise the necessary sum, he had Charlie Allen's approval for the investment in Oil and Minerals.

Theo tried very hard, but he did not succeed. It might be worthwhile to illustrate the complete turnaround in the market. Less than a year earlier I happened to be at a Sunday brunch in New York, where there were several stockbrokers present. One of them, who found out that I was in the oil business, asked me about Planet Oil. This was six months before I first met Theo Ben Nahum. I told this broker that I did not think much of Planet Oil and did not think its shares were worth buying. The man said, "You are crazy! I bought the shares at twenty dollars each and sold them a short time later at eighty. Furthermore, any stock that Charlie Allen brings to the marketplace is a winner. You do not know what you are talking about." He added, "Charlie Allen brought Syntext to the market, the company that introduced the first contraceptive pill. It was one of the greatest successes on the stock market, ever."

The deal with Planet Oil died a slow death. We continued to look for money from many different sources. In the meantime we were conducting an extensive seismic survey offshore in Ghana, another offshore seismic survey in Ethiopia, and we started negotiations for new concessions in Gabon, a rich oil producing country in West Africa. We also applied for a license in Madagascar, one of the largest islands in the world, off the coast of East Africa. But we had still not found a solution to our financial problems.

* * *

I tried another approach which could have assisted us in solving our financial problems and enlarge the scope of our activities. It occurred to me that if we could form a joint venture with Pemex, the Mexican National Oil Company, to explore together in the developing world, especially in Africa, it would greatly benefit both parties. In 1969, Mexico was still seven or eight years away from its major oil discoveries onshore and offshore in the Bay of Campeche and needed additional oil reserves. Needless to say that such cooperation would have assisted us politically as well.

I went to Mexico City and renewed my acquaintance with Jorge Serrano, who was one of the leading Mexican businessmen operating as a drilling contractor. He liked the idea very much. He presented it to

Pemex who supported it. They went to the president of the country to seek approval but were refused. The president stated that he considered foreign oil exploration as a kind of imperialism, and Mexico would not engage in imperialistic activities.

Serrano, several years later, became the head of Pemex.

Chapter Eight

John King and Bernie Kornfeld's IOS Scam

In September 1969, John King wrote a long letter to Dr. Dinstein. In this letter he wrote that the idea of utilizing Israel's good relations in the developing world to obtain oil and mineral rights was an excellent one, whereas the execution of this task by I.N.O.C. was pitiful, and this large hidden potential was going to waste. He therefore suggested that his company, King Resources, take over the concession I.N.O.C. had already assembled and start a vigorous effort to obtain new concessions and explore for oil around the world. King Resources would also marshal the Jewish financial muscle to fund this world-wide effort.

John King was then on top of the world. He was an oilman from Denver who always had big dreams of becoming an oil tycoon. When he associated himself, a couple of years earlier, with Bernard Cornfeld, of IOS fame, who headed the most successful money gathering machine in the world, John King's dreams started to come true. It allowed him to start tapping into the unlimited funds of IOS. In a single year he "milked" IOS of some one hundred and twenty million dollars.

At the time of writing his letter, John King, was using IOS money, to explore onshore and offshore the Sinai peninsula which Israel had seized from Egypt in the Six-Day-War. Not many oil companies were ready to explore in a politically disputed area such as the Sinai peninsula, that was ultimately returned to Egypt eight years later. As a result, John King's standing in government circles was very high.

The facts that John King, within one year, was indicted in the US and served many years in jail, with his company going bankrupt, and that a similar fate awaited Bernie Cornfeld and his funds, and that, furthermore, John King's friend in Israel had later to serve a sentence in an Israeli jail, were still far into the future.

When I saw John King's letter to Dr. Dinstein. I realized that he was very impressed with King's letter. This by itself was quite understandable. King was a rich and famous man of great influence. I saw serious danger looming. I already knew that John King was a crook and that there were very questionable transactions in the IOS history (which I will explain later). There was no use telling this to Dr. Dinstein because he would not have believed me. I saw that he was really tempted to accept John King's offer. For one, it would have solved the problem of the continuous requests for funds for I.N.O.C. voiced by the new influential members of the Board of Directors of the company. It was also possible that Dinstein was under pressure from the Minister Sapir to do something about I.N.O.C. Finally he probably genuinely believed in the vistas opened by John King.

A fluke coincidence helped I.N.O.C. and me personally. Chaim Zadok was about to go to the United States. Zadok, one of the most prominent attorneys in Israel, and one of the leaders of Mapai, the ruling Labor party, was the Minister of Justice in several cabinets and was mentioned from time to time as a future Prime Minister. I knew him well as he was representing Nafta, the other oil company in Israel. Zadok also tried to help me to raise money, several months earlier, by arranging a meeting with Yaacov Meridor, the head of Maritime Fruit Carriers. Zadok knew about our achievements and our financial efforts on Wall Street.

When I heard from him that he was going to the States I suggested that he meet with Lehman Brothers and with one of the prospective investors that I met in New York that summer, a man called Joseph Bonaparte. We had serious discussions with him regarding an investment of five million dollars in our company. (When I first met Bonaparte, I seriously believed that I was dealing with a descendent of Napoleon Bonaparte. When I got to know him better he told me that when his grandfather emigrated to the United States his name was Ben Porat [the son of Porat]. The immigration official did not comprehend this name and he wrote Bonaparte. His grandfather did not object.)

When Zadok returned to Israel he went with me to see Dr. Dinstein. He told Dinstein that he had seen Bonaparte and there was a reasonable chance that five million dollar investment would materialize. He also told him that there was hope that Lehman Brothers will return to deal with our affairs. Dinstein could not ignore Zadok's statements. The danger was over. But it was a close call. I often wondered what would have happened had Dinstein given our assets to King Resources when the company collapsed and John King's indictment and prison sentence followed within a year.

* * *

I first met John King in 1966 when I stayed with Jack Grynberg in Denver, on my first fund-raising trip in the US. Dov Ben Dror, my chairman, and the head of oil affairs in Israel, asked me to see him. Ben Dror was told by one of the leaders of the Jewish community in America that he had been approached by John King with an interesting proposition. King wanted to obtain the entire area of the State of Israel as one oil license. King would then initially raise ten million dollars from the Jewish community in America and conduct exploration in Israel the way it should be done. Ben Dror wanted me to see him and report my impressions.

I met John King in his office on the top floor of the tallest building in Denver. He was tall, large and a very impressive man. He had an enormous office with glass walls all around, from which one could see the Rocky Mountains in all their splendor. His desk, which extended into a boardroom table, was shaped like a V, to allow him to view each participant without any obstructions. Behind his desk was a battery of communication equipment which would not have shamed the command post of an aircraft carrier. It was quite an office and he was quite a character. I told him about our Ziqlag project in Israel. He was not interested at all, but described to me his grandiose plan for exploration in Israel. He said, among other things: "Imagine the Jewish community in America. They would help Israel by investing in my fund, they could write off their trips to Israel as a business expense and they could also find oil and make money. How can you beat this combination."

He told me that he was the first man in America to marry the computer to oil exploration, thereby reducing considerably the odds of drilling dry holes. We went to the basement of his office building where his exploration department was located. We met John King's head of exploration, a professor of geology by the name of Fredricson. Fredricson explained to me how they had succeeded in feeding into the computer all the relevant data derived from the sixty thousand or more wells drilled in Denver Julesburg Basin, near Denver. Their goal was to reach a conclusion as to what combination of sand, porosity, pressure, elevation, and other factors make one well a producer and another well a dry hole.

According to him, they had found the answer. They had reduced the odds from fifteen to one down to five to one. I was extremely impressed.

Several days later there was a convention in Denver of the AAPG (Association of American Petroleum Geologists). In the lobby of the hotel I met a prominent geologist from Tulsa, Dan Bush, who participated in the convention. Dan Bush did consulting work for us in Israel and we were good friends. I asked him whether he knew Professor Fredricson. As Dan Bush was also a professor of geology, I imagined that they would know each other. Bush was noncommittal. I insisted. I said: "Dan, level with me." He then told me that Fredricson was formerly a professor of geology at Pittsburgh University. He bought an island in the Atlantic Ocean for the university, without its knowledge. They quickly fired Fredricson before finding themselves in possession of more real estate in the Atlantic Ocean. I was satisfied that I had found out enough about Fredricson's veracity, or the lack of it. Unfortunately, to this day, more than thirty-five years later, you cannot feed data into the computer and come out with an answer where to find oil.

I found out that something was very wrong in the IOS empire through a chance conversation, on the way to Rome airport, in the summer of 1968. John Christiansen, the geologist, who lived in Rome, was telling us an amusing story. He had sold an oil concession in South West Africa, now Namibia, to a small Canadian company called Bryland Mines, headed by a fellow called Rosenblum. Rosenblum owed Chris twenty-five thousand dollars which he offered to pay in Bryland Mines shares. The shares were then trading at fifty cents a share. Chris refused to take the stock, as he considered both Bryland Mines shares and the Namibian concessions worthless. "Imagine," he told us in the car, "the Bryland Mines stock is now selling for twenty-five dollars a share. I could have had one and a quarter million dollars, but the stock is still worthless," he added.

Several months later an army friend of mine, my neighbor in Zahala, tried to talk me into buying the IOS funds of Bernie Cornfeld. He was truly convinced that this was the way to riches for the small investor. I told him that I did not have any money to invest, but he persisted. As he was a friend and a neighbor, I had to listen to his exhortations about the virtues of IOS. One day he left a brochure of IIT on my table. IIT was the largest fund in IOS's stable, with an eight hundred million dollar portfolio—more than ten billion in today's money. After my friend left, I glanced through the brochure, perusing the list of their investments.

Suddenly I came across the name Bryland Mines. Three hundred and thirty thousand shares, at a cost of thirty-three dollars per share, for a total of almost eleven million dollars. Bryland Mines, I thought, Bryland Mines; where had I heard the name before? And then I remembered Chris's story in Rome. My God, I thought to myself, if this is one rotten investment that I know of, how many more rotten apples are there in their barrel? The man who bought this investment for IOS was not only a fool but a crook as well.

When I was visiting with Jimmy Patino in Geneva in April 1969, discussing our Ethiopian mineral concessions, Jimmy tried to help me with my financing efforts. He was, by the way, the first man who recommended that I should try my best to get to know Lehman Brothers. "They are the best," he said. When Jimmy came to my hotel in the morning to take me to his brother George, he told me that he had just seen Henry Buhl III, who was the number three man in IOS. (IOS's headquarters was in Geneva.) He told Buhl about us and Buhl was ready to invest ten million dollars in our company, the only proviso being that John King, their associate and oil expert, should confirm the value of our assets. "Mr. Patino," I said, "I do not want their money." He looked at me as if I were deranged and asked: "Why not?" I replied: "Because John King is a crook, and I do not trust IOS." Patino probably thought: "This is your own funeral, buddy." And left it at that.

I have often wondered whether my reply to Patino was too hasty or even stupid. If I had accepted IOS's investment, it would have solved all our financial problems and their eventual collapse would not have affected us one iota. On the contrary, we would have had a non-existent investor. But I was so incensed with what I saw and heard about John King, and so disillusioned about the IOS investment in Bryland Mines, that I did not want any part of them. More than a year later Jimmy Patino came especially to London to introduce me to Charles Hambros, the head of the Hambros Merchant Bank. On the way to the Hambros office, he told me that their family's head of investment had asked him: "How did this Alexander know, a year before anybody else, that John King and IOS were no good?"

Chapter Nine

Personal File—The Alexander Family—From Oil to High Tech

I spent two hundred and sixty three days abroad in 1969. All this travel and the various crises left me with very little time to devote to my family. Our children were growing up. Our daughter Shaula finished her army service in 1968 and entered The Hebrew University of Jerusalem in the same year to study mathematics and physics. Our son Kobi was in his last year of high school. I was so engrossed in what I was doing that I hardly noticed their growing up and their passage into adulthood. One would think that such circumstances would have surely caused estrangement in the family. The reverse was true. We were and have remained an outstandingly close-knit family, even though our daughter now lives in New York and our son divides his time between New York and Caesarea. To this day, after each transatlantic telephone conversation—and they call us several times a week—they question each other: How did Dad sound? How did Mother sound? The same questions are repeated on this side of the Atlantic in regard to our children. We judge one another's mood not by what is said but by the inflection in our voices.

Our children were tremendously engrossed in what I was doing. My wife was quite annoyed when my son, while serving in an elite commando unit in the Gaza strip, would call her and his first question was: "Did Dad sign the agreement with Signal yet?" My wife would always complain: "Why does he not tell me first how he is, instead of asking about your affairs?" Our daughter, who went on to become a senior research manager and a CEO, while raising two wonderful children, told my wife several years ago that she had to make a conscious effort to minimize involvement in what I was doing and to focus on her own business and family issues instead.

My wife, Rachel, was less involved in my affairs. She had a lot on her plate, since she was teaching school, raising the children, and taking care of the house and garden in Zahala, where we lived. Luckily, all that did not leave her too much time to worry about my problems.

My wife had been my childhood sweetheart. I first met her at the age of twelve when I came to live with my uncle in the orange-growing colony of Pardess Hanna. I still remember the first time I laid eyes on her. There was this beautiful blond girl performing on stage in a school play. I fell in love with her then and have never stopped loving her since.

I was a newcomer to the country and did not speak Hebrew well. I was not used to living in a farming community, with its relatively primitive conditions, coming, as I did, from the big city of Warsaw. My widowed mother brought me from Poland to Palestine in 1934. My father died in Poland when I was nine years old, and my only brother, Sioma, who was studying at Mikveh Israel in Palestine, died the same year. I was a very difficult child, so my mother, in order to be able to organize her life in the new country, sent me to stay with my uncle in Pardess Hanna.

My wife, on the other hand, was a native Israeli, born and raised there. Her family were true pioneers, comparable to the pilgrim Fathers who had come to America on the Mayflower. Her uncle, Shlomo Yaacov Chazanov, came to Palestine in 1882 with a group of settlers from Russia. They belonged to a group called BILU, which was an abbreviation of a sentence in the Bible, "House of Jacob, come ye and let us go" (to the Land of Israel). In 1884, ten of them established one of the first agricultural colonies in the country, which they named Gedera.

My wife was a good athlete, which I was not. She was all the things I wanted to be. Although I spent almost a whole year in Pardess Hanna and continued to return there for most of my summer vacations, staying with my uncle's family, I do not think that my future wife and I exchanged too many words before the age of sixteen. At that time Rachel was accepted at the Jerusalem Gymnasia (High School). Very few youngsters went to high school in those years. I believe that she was the only one of her class in Pardess Hanna to continue her studies in high school. As she was lacking the first two years of high school (her older sister, Heftzibah, had to complete high school studies before the family could afford to send Rachel there), I was asked by my cousin, who had in the meantime married Rachel's brother, to teach her mathematics and physics. These two subjects in my high school studies at Gymnasia Herzliya in Tel Aviv, I knew well and liked. We spent the whole summer in a "sukkah" (hut) formed by a mulberry tree in their garden, going over those two subjects.

Apparently my teaching was adequate as Rachel was accepted to the 11th grade in the Jerusalem Gymnasia without any difficulty. Perhaps we touched hands once during this long summer.

We started going out more or less steadily after we graduated from our respective high schools. But she always had several other boys courting her at the same time. Her motto was, the more the better. Let them fight and be jealous. She later tried to convince our daughter of the advantages of this system, but she never succeeded. Our daughter is super-conservative, but mother and daughter are also the best of friends, in the manner of two high school girl pals.

My wife went to a teachers' college and I joined the British Army. Part of the time during my service we were going steady, part of the time she had other boyfriends. Finally, in my last year of service, when I was stationed in Egypt, I received a surprise letter from Rachel saying she had chosen to marry me, and wanted to get married soon. We decided to have our wedding during the Passover vacation in 1946. As there was a postal strike in Palestine, I could not find out the date of the marriage ceremony. I arrived in Tel Aviv from Egypt one morning, but my mother was not at home. I went to have a haircut at the barbershop that I always frequented, that was near my mother's flat. There were several people waiting. The barber told them: "Let me take care of this young man first. He is getting married today." That is how I found out my wedding date.

We have been married ever since, and I do not think that a person can be blessed with a more wonderful wife. Ray Christian, my partner in Oklahoma, told me one night, after he had had a little too much to drink: "You are working too hard. You are going to die soon. Then I am going to marry Rachel." All my friends respect her unusual wisdom and broad knowledge in so many areas.

Our two children are very different from each other. They fought incessantly, like cat and dog, until their late teens. They now love, and what is more important, respect each other immensely. Our daughter Shaula is very talented in many spheres and very conservative in her views and behavior. She holds a Ph.D. in computer science, and has had various careers. After completing her Ph.D., she was an assistant professor at NYU, then she moved to IBM Research, where she was Senior Manager for Distributed Systems Software Technology and received an IBM Outstanding Innovation Award. She left IBM to found SMARTS, where today she is the CEO. Shaula is multilingual, very attractive, and excels in anything she is interested in, and this covers many subjects—except sports. Shaula serves as the Alexander family chief medical officer—she is first to be consulted on any medical issue. Shaula was named after my wife's nephew, Shaul Pnueli who was killed during the War of Independence, together with thirty-four of his comrades on their way to the beleaguered Gush Etzion settlements in the Hebron hills, carrying plasma and ammunition. They walked more than twenty miles during the night and were ultimately discovered at daybreak and attacked by thousands of Arabs. They fought valiantly for eight hours until none of them was left alive. This group of thirty-five heroes—the 'Lamed Heh'—and their last battle have become a legend.

Like all Israeli high school graduates, Shaula served in the army. She was posted to Air Force Intelligence, and during the Six-Day-War she served in the aerial photography interpretation unit. On the first day of the war, there was a total blackout, and we spent part of the evening in a trench that we had dug in our garden, as the Jordanian Army was shelling our neighborhood from the nearby Arab town of Qalqiliya. Shaula returned home late at night, and called to us to get out of the trench. "The war is practically over. I myself counted more than two hundred Egyptian planes destroyed by our Air Force in the first four hours of the war," she said.

When Shaula married, she kept her maiden name. Her ex-husband, Yehiam Yemini, whom she met at The Hebrew University, served in the most elite unit of the Israeli Army. Whenever one reads about daring exploits of the Israeli Army, including the freeing of the hostages at Entebbe, they were carried out by this unit. Yehiam is a professor of computer science at Columbia University in New York. Shaula and Yehiam have two lovely children—Eviatar and Noa.

In 1993 Shaula left IBM, to found a software company called SMARTS (System Management Arts) with her former husband. SMARTS has developed and patented the first and only technology to automatically analyze dynamic distributed systems, enabling its software to manage the world's most complex information technology (IT) systems.

Most of the world's largest telecommunications companies, as well as many of its global enterprises, rely on SMARTS software to assure delivery of IT-dependent business. In spite of the tough economic environment, SMARTS' revenues and profits have continued to grow in recent years. SMARTS is privately held and has two outside investors: George Soros Funds and Bessemer Venture Partners.

Our son Jacob, who was referred to as Kobi from birth and would probably not respond to being called Jacob, was a born athlete. He was the captain of the handball and other sports teams at school. When we lived in New York in the 1950s he was always elected president of his class, from kindergarten on. When I commented on it he cried, "I'm too tired to be president all the time."

Kobi went to the same high school that I had attended, Gymnasia Herzliya in Tel Aviv. He too served in an elite commando unit, and I am proud of this. It is the custom among the choicest individuals of Israel's youth, especially those born in a kibbutz or in a wealthy and privileged suburb, to volunteer for the most daring units, such as the commandos, paratroopers or naval commandos—this was not the case when I grew up. After two years of service, Kobi was sent to an officers' course, followed by a course for intelligence officers, and subsequently he served as a deputy intelligence officer of the Jerusalem Brigade at the time of the Yom Kippur War. But this story will be told later.

After his army service Kobi studied economics at The Hebrew University, from which he graduated *cum laude*. He then went to New York, where he worked during the day for Shearson Loeb Rhoades (that later became Salmon Smith Barney), the large New York investment house. At night he attended New York University, where he received an M.B.A. two years later. He was offered admission into the M.B.A. program at Harvard Business School and, to this day, I cannot understand why he did not grasp this opportunity.

Kobi wanted very much to work with me as he was fascinated by the oil business and the various crazy ventures that I sometimes got involved in. The best thing that could ever happen to him occurred in 1982. In that year I lost almost all my savings following one investment that my wife called a speculation, and she was probably right. As a result, there was no more room for both of us in what I was trying to do. He had to make it on his own and he did so in a very big way.

In the same year, 1982, Kobi got together with his brother-in-law Yehiam Yemini, and a computer engineer called Boaz Misholi. They formed a voice-mail company that was first named Efrat and then renamed Comverse Technology.

For the first ten years the company struggled to survive and all of Kobi's time and efforts went into raising the funds to keep it alive. His two partners left the company during that period. In 1987 Kobi became chairman, president and CEO of the company. Since then, he has succeeded beyond expectations. Comverse today is one of the three largest high technology companies associated with Israel, employing more than four thousand people as of 2003. Half of them work in Israel and the other half in the US and other parts of the world.

Comverse Technology Inc. is the world's leading provider of software and systems enabling network-based multimedia-enhanced communication services. More than four hundred wireless and wireline tele-communication network operators, in more than one hundred countries, have selected Comverse's enhanced services systems and software, which enables the provision of revenue generating value-added services. Comverse Technology is an S & P 500 company and NASDAQ-100 index company. Comverse and its two subsidiaries, Verint Systems and Ulticom, trade on the NASDAQ Exchange. Comverse has more than two billion dollars in cash, probably one of the largest hoards of cash for a company of its size in the world.

A large part of Kobi's success may be attributed to the three strategic decisions he made when he took over leadership of the company. They were: First, to sell only to telephone companies; they are the largest consumers of voice-messaging services, and when they buy, they buy big. Second, to focus on the international market which was virgin territory, while the US market was very competitive. Therefore, Comverse was able to become the international leader and then, in the late 90s, it re-entered the US market and became the world leader in enhanced services. The third decision was to raise money whenever possible, and as much as possible. This is the reason for the accumulation of cash that the company possesses and its consequent financial strength.

The mobile telephone explosion which occurred during that period obviously helped in a major way because it more than doubled the number of telephone companies in the world.

Due to the great difficulties experienced in establishing Comverse and in keeping it alive in its early years, which required continuous travel between Israel and the US, Kobi married quite late. His wife, Hana, is ten years younger than he. When he introduces her, he says: "This is my second wife—I skipped the first wife," and indeed she looks like a second wife.

Almost all of Kobi's childhood friends are divorced or married for a second time, while Kobi's marriage is a very happy one. They have three charming children. There is nothing that can spoil the good mood of Hana. Kobi says that even if a glass is almost empty, it is full of nectar to her.

Chapter Ten

Kewanee Oil—Ghana

The terms of our concession in Ghana called for the completion of the seismic survey during 1969 and the drilling of at least 12,000 feet during the following eighteen months. We therefore had to start a seismic survey in the first half of 1969. As six of our eight blocks were offshore it was going to be a large and costly survey.

Ray Christian, our partner, financed his 46 percent of the cost from the proceeds of the partial sale of his interest to three bankers in New York. We financed our share partly by using accumulated operators' fees and partly by borrowing funds from our Ethiopian joint venture.

The seismic survey revealed the existence of numerous prospective features in our offshore areas, the most prominent of which were a large structure in the Keta Basin in the eastern part of Ghana and two structures in the western part of Ghana in the Tano Basin. We named the larger structure in Tano as Tano, and the second structure, in deeper waters Tano South.

In fact, the seismic results were so encouraging that when I showed them to Mark Millard, the banker from Loeb Rhoades, he got so excited and keen to get involved in our Ghanaian venture, that he offered me the use of his sports car. Its special attraction, he explained, was that it had six forward shifts. Regretfully, I decided to decline his offer as there was not much I could do with a sports car in New York City.

* * *

During the same period in the summer of 1969, I saw Max Fisher—one of the foremost leaders of the Jewish community in the USA. Fisher was a very wealthy investor and the largest private stockholder in Marathon Oil.

As I have mentioned previously, he represented Signal Oil & Gas on the Board of Directors of Paz Oil in Israel. He was the man who discouraged Meir Sherman, the president of Paz, from supporting my efforts in foreign oil exploration. But this was in 1967, two years earlier. When Max Fisher saw what we had accomplished in the intervening two years, he immediately sent an associate of his, an oilman by the name of Harold McClure, to meet me in New York and review our holdings and our plans. Fisher then offered to finance one hundred percent of all our future expenses for a half interest in our holdings

During the conversation with McClure, I realized that they wanted not only 50 percent of our future concessions, which we would acquire with their financial assistance, but also 50 percent of what we had already assembled. It meant, among other things, the contribution, for free, of half of our Ghana holdings that the market was appraising at ten million dollars. I considered that proposal unfair and declined their offer.

I have often wondered whether I made a wise decision. That is was a just decision I have no doubt. But, in retrospect, it was probably a rash and unwise decision, especially from my personal point of view.

In one fell swoop, we could have eliminated all our financial worries and would have been able to devote all our efforts to negotiating for new areas around the world. We would also have solved the problem of searching for large industry partners. Max Fisher and his associates, among whom he counted Henry Ford, offered to perform this function for the proposed joint venture. On a personal level, the agreement with Max Fisher would have immediately solved my financial worries and my salary problem. My salary as a government employee, no matter how high the position was, did not suffice for a decent standard of living. My wife had to continue teaching at her school, so that we would have additional income. Only our combined salaries allowed us to pay tuition fees for our son in high school and provide the means for our daughter to go to university after her army service.

Another immediate benefit of making a deal with Max Fisher would have been the removal, once and for all, of the constant threat to the independence of I.N.O.C. Max Fisher was so powerful on the Israeli scene that nobody would have dared to touch us. For many years I was afraid that one day Minister Sapir would accept the advice of my "friends" in the industry and terminate our independent existence. This was not an empty threat. Dr. Dinstein voiced it on several occasions, stating that the only way he managed to prevent our forced merger with Lapidoth, my former employer, was by assuring Sapir that we did not appear in the government budget and did not cost the government any money.

But I made that rash decision, regarding the joint venture with Max Fisher, and our contacts terminated at that juncture.

During the same period of the summer of 1969, I had another interesting offer, from the chairman of US Smelting and Refining. He suggested that we pool our acreage positions—their large holdings onshore USA and our foreign holdings—into one company which US Smelting would finance from then onward. The problem was his demand that each party contribute its holdings at cost to the new company. I tried to argue this was not a fair proposition. Our concessions were acquired at almost no cost, whereas his leases in the USA were purchased at full cost. I could not convince him.

Several months later we were negotiating with Joseph Bonaparte, who was contemplating a five million dollar investment in our company. Bonaparte offered to revive our contacts with US Smelting. He claimed that I was too hot-headed. Do not say, "No," say "Yes, but…" His contention was that a reasonable agreement could have been reached with US Smelting and, as in the case with Max Fisher, all our financial worries would have been over. Six to eight years later, I read that US Smelting was under attack from a corporate raider and, in order to get rid of him, they went into voluntary liquidation. The sale of their assets brought around four hundred million dollars. When I read this article, I realized that we could have owned 50 percent of a major corporation and, who knows, perhaps we could have prevented its liquidation.

*　*　*

When we received the maps with the seismic interpretation of our Ghanaian offshore areas, the time came to start looking for drilling partners with "big dollars." There was a lot of excitement in the oil industry regarding the oil prospects of West Africa. It was a virgin area as far as oil exploration was concerned. Furthermore, the exploration of Ghana followed the exciting recent discoveries offshore Nigeria. The discoveries offshore Indonesia also added fuel to the offshore fever. We therefore did not anticipate a serious problem in finding rich partners. We did not have too much time, as we had to start drilling by July 1, 1970.

The mobilization of an offshore drilling rig was going to take three to four months. We also had to prepare a land base and the necessary drilling supplies, which, by itself was quite a logistic feat. Both Ray Christian and our company therefore had to conclude a "farm out" agreement with large companies around the end of 1969. We decided between ourselves that each of us would seek his own partner.

Ray Christian's Mayflower company farmed out part of its interest to Hamilton Brothers Oil Company, of Denver, Colorado. Hamilton was a very successful, independent oil company, with substantial production in the US and the North Sea. They had undertaken to fund all Mayflower's share (46 percent) of the cost of the two wells, one in the east and one in the west of Ghana, for a 20 percent interest. Mayflower group (including the bankers who funded the seismic survey) would remain the owner of 26 percent at no cost.

At the same time, we talked to several prospective companies. A serious dialogue developed with the Kewanee Oil Company, from Bryn Mawr, Pennsylvania. Kewanee was the oldest oil company in the US, founded in 1871 in Pennsylvania, where the first oil in the United States was discovered. At the time of our negotiations, they had already been in business for ninety-nine years.

The crucial meeting was in their offices in Bryn Mawr in late February 1970. It was a very cold and snowy morning. I remember sitting by myself on one side of a table, with five vice-presidents of Kewanee on the other side, including their executive vice-president. His name was Boland, and he was actually the acting president of the company. The president, a Mr. Smith, whose grandfather had founded the company, was much more interested in growing tomatoes in the Bahamas than in the oil business. Smith very seldom visited his company offices in the US.

I felt very uncomfortable at this meeting. In addition to feeling overpowered by the other side, one of my front teeth was wobbly and I was afraid that I might swallow it by accident. I was therefore not in the best shape for this crucial meeting. I vowed to myself then that when I would have enough money, I would always have somebody accompany me to important meetings.

I tried to achieve two separate goals at that meeting: funding for the drilling of the two wells in Ghana and an equity investment by Kewanee in I.N.O.C. As described previously, I needed an equity investment by an oil company. An oil company shareholder in I.N.O.C. would attract non-oil investors to buy shares in our company. The Kewanee executives did not understand my insistence on an equity investment. They were interested in Ghana and were ready to pay whatever it took to explore in our blocks there. But they were not interested in me or in my company. It is much clearer to me now than it was then.

During the meeting I offered numerous scenarios, all of which were based on a formula of dividing Kewanee's contribution to the drilling of the two wells into two channels, one being a direct investment in the drilling expense and the other a stock investment in our company.

We concluded the meeting with my proposal that they would invest 1.7 million dollars, which covered 54 percent of the projected cost of drilling of the two wells in Ghana, for a 15 percent interest in the licensed area, while we would remain with 39 percent; and that they would also invest two hundred thousand dollars in the shares of our company. When I left, I had a feeling that they would agree to those terms. We decided to be in telephone contact the following day, when Kewanee was going to advise me of their decision.

The next day I waited for Kewanee's call in Howard Marr's office in New York. Boland, the acting president, phoned, and we discussed the terms again. Once more, I insisted on an equity investment, and he promised to call back after an hour. Howard Marr, who was listening in on our conversation, said: "You were very good. Continue to be firm. You are winning."

The phone rang again. Boland was on the line. "Mr. Alexander," he said, "we have examined the transaction from various angles, and I do not think that we can reach an agreement. Thank you. Good-bye."

I was shattered. My whole world had collapsed. It was almost the month of March. Several days earlier, Ray Christian and I were discussing the ordering of the drilling ship, I told him to go ahead with it. He warned: "Look, if you do not come up with your 54 percent in time, we will go bankrupt."

"We will come up with our share," I promised.

Now, here I was with my deal disappearing. Not only was there danger of irreparable damage to my friend and partner, Ray Christian, but we might also lose our concession in Ghana, our crown jewel.

No matter how hard I tried to understand what had happened, I was completely mystified. It looked like they had agreed, and then, suddenly they withdrew. I called Howard Marshall—the grand old man of the oil industry and my partner in Nigeria—in Houston. I told him the story and asked for help, advice, and possibly intervention. He did not know anybody in Kewanee and could not help. I examined, with Howard Marr, every phase of our negotiations but could not discern where things had gone wrong. Finally, I called Lehman Brothers.

Don Russell was travelling abroad. I asked to speak to Ed James, who was Russell's associate in Lehman's oil division. I explained the situation to him. He said that, as he was not dealing with our affairs and did not know the particulars of our negotiations, we should wait until Don Russell returned. I begged him to intervene and explained that the fate of our company depended on it. He finally promised to check with Kewanee as to what happened and to call me back.

An hour later he returned my call. "Kewanee is interested in the deal. For an investment of 1.7 million dollars for a 15 percent interest, as discussed, you can conclude the transaction. You wanted an additional two hundred thousand dollars equity investment; that was too much. At a 1.7 million dollar figure they are interested."

I was speechless, could not believe my ears. "What should I do next?"

"Can you be here at 2 o'clock?" he asked.

"I will be there."

Howard Marr accompanied me and we arrived at Ed James' office at 2 o'clock. He repeated to us the conversation he had had with Kewanee. I said: "Look, we do not have a penny. All the 1.7 million that they contribute will go towards drilling the wells. Perhaps you can get some cash for us?"

"How much do you want?" he asked.

"One hundred thousand dollars," I said.

"Let us try for eighty thousand," he suggested, "it sounds nicer."

Ed called Kewanee and asked for Don Hockaday, the vice-president in charge of exploration. "Don," he said, "Mr. Alexander was gracious enough to come to our office (gracious enough! I was ready to crawl on my belly all the way from 50th Street to Wall Street). Mr. Alexander agrees to the 1.7 million deal, but he very much needs some cash, as well."

"How much?" Don asked.

"Eighty thousand dollars," Ed said.

"Fine," replied Hockaday. Our company was saved.

This was on a Thursday. The following Sunday, Adi Ephrat, our attorney, arrived from Israel to conclude the contract with Kewanee. On Monday, we all met in Ken Bialkin's office and, by that night, a twenty-five page "farm out" agreement was signed with Kewanee. The following day, as there was a Western Union strike, a special messenger from Kewanee delivered five hundred thousand dollars to our bank account in New York.

Strangely enough, although this was our biggest achievement to date, when we finally became a real company, with money in the bank, reputable industry partners, and two expensive offshore wells about to be drilled, at no cost to ourselves, I felt deflated rather than elated. I felt that I had not reached my original goal. We did not yet have an equity investment and could not start building a solid company. When I arrived in Israel everybody, from the Justice Minister Zadok down, congratulated me but I felt very strange.

I learned several important lessons from this traumatic experience. The first was that, although I always recognized that there was a wide cultural gap between Israelis and gentile Americans, I had believed that I personally was able to breach that gap. As a matter of fact, however, I probably still suffered from this cultural diversity. We Israelis are used to overstatements, whereas most Americans, especially those I met in the southwest, speak in understatements. Although some of us speak very good English, there is still a huge language gap between the two sides. For example, when an American says "I understand your point of view," he probably means that you are talking nonsense but he does not want to offend you. An Israeli, on the other hand, interprets such a statement to mean that the American agrees with his point of view. When, in later years, we had a large American company as an investor and many large American companies as our partners around the world, I often received reports from our executives that they had come to an agreement with our partners. Only later did I find out that they had merely been told that their point of view was understood. No agreement had been reached or would be forthcoming.

The second lesson was that Americans do not usually haggle. If you mention a figure, they take it as a well-thought-out position, and they either agree or do not accept the deal. To many people of other nationalities, for whom arguing about a price is an essential part of any commercial transaction and to whom the price mentioned is only the opening gambit, this American custom is difficult to comprehend.

When I next met Don Russell (of Lehman Brothers), he suggested that we pay them a fee of fifty thousand dollars for the Kewanee transaction. I was incensed! We had been dealing with Lehman Brothers for almost a year. They had caused us more grief than good. They had several times withdrawn from dealing with our financing. They did not get us one red cent of investment. And then, when finally they did something productive for us, they wanted all of fifty thousand dollars.

Moreover, fifty thousand dollars would have wiped out almost all the cash contributions from Kewanee, leaving us penniless again.

I tried to explain all of this to Don Russell. Several days later, Don Russell informed me that Lehman Brothers did not wish to be associated any longer with our financing effort. When Chris, our friend from Rome, heard this story from me several years later, he told me that I had made a terrible mistake. I should have paid them, even if I had to beg, borrow or steal the money. Chris was of course right. Retaining the Lehman name and their support was of immense importance, and I had lost it.

* * *

I often bragged that we were the only company that did not cost its stockholders (the Israeli government) any money. But what a high price was paid for this doubtful achievement. What an impossible situation our government had put us in. They did not understand or appreciate the fact that this was the first time "in two thousand years" that an Israeli oil company was sitting at the negotiating table with major oil companies as equal partners, discussing joint exploration projects. This was achieved in spite of the Arab boycott. Israeli government policy, instead of providing us with the necessary funds to participate in exploration with our distinguished partners, forced us into a position where we had to sell to our prospective partners segments of our concessions and charge them high overhead and operator fees to boot. We had to do it in order to survive and be able to pay our salaries and travel expenses. The non-payment to Lehman would never have occurred had we had sufficient money, at least for day-to-day expenses.

Chapter Eleven

The Rothschilds

I started writing this book in 1988 and completed the first nine chapters. In 1989, I became very busy with new exploration ventures in the Far East and Russia, and only ten years later did I resume writing.

The man to whom I owe the continuation of writing this book is my friend Hezi Carmel, a writer and journalist, and a former Mossad representative in France.

When we met one day in Tel Aviv, in 1998, I told him about the book and asked his advice as to whether I should persevere and whether it might have general appeal.

A week later, when we were already back in London, I received a call from Hezi telling me that I owe him a whole night's sleep. He started reading the manuscript when he went to bed and could not put it down for six hours. When he finished at four in the morning, he wondered: "What happened next?" He strongly urged me to continue and finish writing my story.

Rereading the first nine chapters I realize that it is very difficult for a reader today to grasp the feeling of isolation and frustration that an Israeli oilman felt in the 1960s and 1970s. It is very hard to comprehend today that to have the involvement of Lehman Brothers in an Israeli deal was considered to be the height of business and political achievement.

In the summer of 1969, my wife and I were having dinner at the home of my chairman and his wife, the Ben Drors. There were several heads of Israel's economy there. Ernest Japhet, the chairman of Bank Leumi, the largest bank in Israel, was asking me how my financing efforts were succeeding. Ben Dror, who was standing behind us and heard Japhet's question, said: "If Zvi succeeds, he will change the GNP (gross national product) of the State of Israel."

Today, Lehman Brothers and every other investment house on Wall Street are involved in Israel and raise money for various Israeli enterprise. The same Lehman Brothers were the lead underwriters for several public issues of my son's company, Comverse Technology, for which they raised more than a billion dollars. The last effort was in May 2003, when they raised four hundred million dollars for convertible debentures, without any interest to be paid by Comverse. This large sum was raised overnight, between 4 p.m. and 8 a.m. the next morning—without any prospectus.

* * *

In late 1969, we approached many other potential investors. One of them was Bank Leumi, mentioned above. We met Ernest Japhet several times, had long discussions, but nothing came of it.

In February 1970, following Dr. Dinstein's introduction I met with Sir Sigmund Warburg, one of the most prestigious Jewish bankers in the world—in the same league as the Rothschilds. The House of Warburg goes back hundreds of years. Sigmund Warburg left Germany in 1939 and came to England, where he succeeded, in a short time, in establishing one of the most important and prestigious merchant banks in Europe—the House of S. G. Warburg. I was warned by Zevi Dinstein to be ten minutes early, as Sir Sigmund, being of German origin, was very punctual. Needless to say, I was twenty minutes early for our meeting in the Tel Aviv Hilton. I told Warburg what we were doing, gave him a file of our projects. But I did not hear from him again.

Another funding effort was with Baron Edmond de Rothschild. Theo Ben Nahum, whom I mentioned in regard to Planet Oil, was personally very close to Edmond and had looked after him for a period of time when Edmond's parents divorced. Theo phoned Edmond in Paris from New York, and told him that they—Planet Oil—were going to invest one and a half million dollars in our company. He invited Rothschild to participate, and to invest an additional sum of one and a half million. Edmond replied that he would consider it favorably, and invited Theo to come to Paris to discuss it. We went together to Paris to meet with the Rothschild team. Theo went to see Edmond, and I was asked to meet with Bollack, who was one of Edmond Rothschild's most senior executives.

Bollack knew Israel well, as he was the president of the first Israeli oil pipeline, from Eilat to Ashkelon, which was financed and managed by

Edmond de Rothschild's interests. This "small pipeline" was built after the Sinai campaign of 1956, when the Egyptian blockade of the Straits of Tiran was removed and passage to Eilat, through the Red Sea, was reopened.

Bollack received me very graciously and, after hearing what we were doing, became enthusiastic. He discussed our future plans with me in detail and assured me that the House of Rothschild was going to invest in, and support all our efforts. But I did not hear from him again after that meeting.

I believe that I found out subsequently why there was silence from Warburg and Rothschild, in spite of the very good initial meetings. It was most probably due to the attitude of the Nahamias family in Paris.

The Nahamias family was a very prominent and extremely wealthy family of Jewish shipowners and oilmen living in Paris. Joseph Nahamias, the oldest brother, was involved in several Israeli enterprises as a director and as an advisor to the Israeli government. A famous story going around at the time was that Prime Minister Eskhol asked Joseph Nahamias to be an advisor to the Israeli government. Nahamias replied: "Mr. Prime Minister, it will cost you a thousand dollars a day" (a very large sum at the time).

"But, Mr. Nahamias, we are a very poor country, and you do not really need the money," remonstrated Eshkol.

Nahamias replied: "Mr. Prime Minister, the value of the advice is exactly in accordance with what you pay for it." Eshkol paid.

I met Joseph Nahamias in our Tel Aviv office, and we had a very pleasant meeting. At his suggestion, I went to see his brother, who was in charge of their family's oil affairs, at the Pierre Hotel in New York. The latter's attitude to our proposition and to our concessions in Africa was very cool. He said that he believed only in Canada, and that they were investing all their exploration effort there. I told him that there was no comparison between the size of our holdings in Africa and those that he could obtain in Canada; after all, everything else being equal, the discovery of oil is to some extent related to the dimensions of one's prospective acreage. I felt that I did not convince him.

Don Russell, of Lehman Brothers, also phoned Nahamias on our behalf. Russell himself was a director, and a large shareholder, of a small Canadian oil company, Sunlite Oil. Russell told Nahamias that Canada and Africa were two entirely different creatures. Both areas had their specific merits, and the fact that we enjoyed a unique standing in Africa was of great economic importance. Russell probably did not convince Nahamias either.

What most probably happened was that both Sigmund Warburg and Baron Edmond de Rothschild invited Nahamias to participate with them in investing in our company. They must have also asked his opinion as to the merits of such an investment, as Nahamias was the oil expert among them. When Nahamias declined and refused to invest, Warburg and Rothschild got cold feet. Theo Ben Nahum later received a confirmation from Bollack of my suspicions. Bollack told Ben Nahum, after our meeting, that he was extremely impressed and that our company was the most exciting investment opportunity he had come across recently. He said that he was going to recommend it strongly and that the House of Rothschild was definitely going to invest with us. When Theo Ben Nahum inquired subsequently, on numerous occasions, what was happening with the Rothschild investment he always received the reply that they were still considering it. That is another polite way of saying no.

In a way, I can understand the Rothschild and Warburg decisions. To invest millions of dollars in a new company which does not have any oil production yet, you really have to be convinced and believe in the prospects of the company and in its management. If one of your group, who is also the only expert amongst you, declines, the easiest decision is to say no.

I had one more experience with the House of Rothschild, this time with the English branch. In the summer of 1970, after Jimmy Patino introduced me to Charles Hambros, the head of the prominent Hambros Merchant Bank, the latter suggested I should meet the Rothschilds in London. I inquired about the composition of the Rothschild bank and found the highest official to be a man called Rodney Leach. I was very glad that he was not Jewish because I wanted, from the outset, to establish a pure business environment. I arranged an appointment with Rodney Leach and found a sympathetic ear from the outset. He spent almost three hours with me. I was already experienced enough by then to detect, in the first fifteen minutes of a meeting with a potential investor, whether we were making contact and whether the chemistry worked. You can see it in the eyes of your interlocutor, in the nature of his questions, and in the general ambience. Rodney Leach was extremely interested. After the first half hour, he brought in an assistant to take notes of the discussion. One of Leach's observations was: "You have a very important portfolio of assets, and it is of great interest." By that time we had had two more concessions, one in Gabon, an oil producing country, and another very large concession on the island of Madagascar.

Before we parted, I told him that he would do us a great favor if he presented the venture to his principals, Jacob and Evelyn de Rothschild, as a pure oil venture and not an Israeli or Jewish related investment. He promised to do so. When he asked how I was going back, I said that I was taking a taxi. He said, "No you don't." Instead, he took me down to where the Rothschild cars were parked, with the Rothschild crest painted on them, and requested one of the drivers to take me to my destination.

Rodney Leach promised me a reply within a week. Sure enough, a week later he telephoned me in Tel Aviv and we spoke for almost half an hour. He talked with great sorrow and anger. He said that he presented the project to Jacob Rothschild and the rest of the Rothschild family. The consensus of opinion was that, as the Rothschild family was so deeply involved in philanthropic activities in Israel, their involvement in a business venture would create a "credibility gap" with the prospective investors they might approach. As a result, they chose not to be involved in our affairs. I said: "Mr. Leach, in other words, if I came to you with the same portfolio of assets representing the National Oil Company of Nicaragua, the House of Rothschild would be ready to finance it." He replied: "Definitely so, Mr. Alexander." I believe that now, almost thirty years later, Rodney Leach still remembers this conversation. He was incensed by the injustice in the Rothschild's reply and in what he considered a lost opportunity. I submitted the text of this conversation to Dr. Dinstein. His reply was that it was the Rothschilds' prerogative how to address such a situation, and that was that.

There were other efforts to obtain finance, but they were of lesser importance, and none of them was successful.

Chapter Twelve

"Payments" in Gabon

In mid-1969, we applied for several licenses offshore Gabon. Gabon, a very large country in West Africa, is well endowed with natural resources—timber, oil, and other minerals. The population is tiny, less than half a million people. As Gabon was formerly a French colony, all the oil licenses belonged to Elf—the French government-owned company. As in many other countries, a relinquishment system was in practice. Such a system initially allots large oil licenses to an oil exploration company for a specific period of time, usually three to five years. At the end of the initial period, the company has to choose its preferred areas and relinquish back to the Government the remainder of the area—usually 50 percent of the original. By such a procedure a larger region is initially studied and covered by various surveys, many times including expensive seismic surveys as well. The Government then gets back an area with added value, which it can then allot to other explorers at a higher premium. A second relinquishment period follows, usually two years later, when the original explorer has to relinquish another 50 percent of his remaining area, and the procedure of allotting that second relinquished area is repeated.

If and when an oil discovery is made, and then declared to be a commercial discovery, an area surrounding the oil field is granted to the discoverer as a long term lease, to develop and produce for a period of twenty years or more, depending on the prevailing oil laws and regulations of the country.

As Elf had been allotted, and relinquished, numerous areas offshore Gabon, we applied for five such relinquished blocks.

Israeli relations with Gabon were excellent, and the Israeli ambassador kept assuring us that a positive reply was forthcoming. When almost a year had passed and nothing had happened, we almost lost hope of being granted a license in Gabon.

I mentioned previously that we had a very good relationship with the Israeli Ministry for Foreign Affairs. There was only one bone of contention between us. The ministry voiced its disapproval of "payments" to Government officials in Africa. The fact that this was par for the course in all of Africa, as I described in the chapter dealing with Nigeria, was not accepted by them.

One day, we had a long meeting in the ministry regarding such "payments," where our divergent views were expressed. Approximately two weeks after that meeting, an urgent cable arrived in the ministry from a high official of the Israeli United Nation Mission in New York. The telegram stated that he had met with an old acquaintance from his African service days. That gentleman informed him that an Israeli company (meaning us) had applied for oil exploration blocks in Gabon. The man added that if this company was serious and really wanted to receive the exploration blocks, it had to come to an appropriate agreement with the relevant authorities in Gabon. Furthermore, the cable emphasized that this information should not be divulged to the Israeli ambassador in Gabon under any circumstances. All negotiations henceforth must be conducted only between I.N.O.C. and this Gabonese emissary.

The next day, I was urgently called to the Foreign Affairs Ministry in Jerusalem. The meeting lasted several hours, and it was quite amusing. I was first asked for my reaction to the telegram. I replied that, as the telegram was directed to the ministry, they should first address the subject. I was asked whether the areas were very desirable. I replied in the affirmative, as Gabon was one of the richest oil-producing countries in Africa. Then I was asked again for my opinion. I replied that they, the government ministry, were my shareholders; therefore they had to set policy, especially as only two weeks earlier they had expressed very clearly their policy against such unauthorized payments.

They hemmed and hawed, but I did not let them off the hook. I insisted that they decide. After several hours of back and forth discussions, I was instructed to go ahead, to make the necessary arrangements, but to be extremely secretive and careful.

Several weeks later, I met our high ranking UN official in New York. He introduced me to his African friend. This friend had probably made enough money already, as he was living on the French Riviera. He was not even a Gabonese citizen. Twenty-eight years later, in 1998, I heard his name mentioned as one of the leading ministers in the new Congolese government.

I had lengthy discussions with this African gentleman regarding the terms of our agreement. We finally agreed that most of the payments would be staged following our work program; thus the bulk of the payments would be delayed for several years. Only a small sum would be paid upon the granting of the licenses. Additional sums would be paid after the completion of the seismic surveys, and more payments at the commencement of drilling. Obviously, far more substantial sums would be remitted after discovery of a commercial oil field.

Soon thereafter, our ambassador in Gabon advised us joyfully that his long efforts had borne fruit, and we were going to be allocated our desired licenses.

A few weeks before the award of the licenses, we received, by regular mail, a telegram from our African friend on the French Riviera, asking for very substantial immediate payments. These had not been discussed previously. I phoned him right away and stressed strongly that he should never communicate with me by open cables. Secondly, I advised him that we must either stick to the original terms or forget about the whole deal. He finally agreed to both demands.

I sent our lawyer, Micki Deouell, to Geneva to make the initial payment, with instructions that the funds be transferred from our bank to our lawyer's bank account in Geneva, and from there to the recipient's bank account. I remembered Dov Ben Dror's kitchen analogy, that when you transfer cooking oil from one container to another, although the entire quantity is being transferred, your hands somehow become oily. I decided that, in our dealings, our people's hands should remain completely devoid of any oil coating.

Several years later I read that Ashland Oil and its chairman, Oren Atkins, whom I mentioned in our Nigerian saga, were accused by the American Government of making multi-million dollar "unauthorized" payments in Gabon. Atkins, I believe, was indicted for this "crime." The American government is very sanctimonious in this respect and refuses to accept the facts of life in the developing world. Many American companies suffer from this law and have to devise various circuitous ways in order to live in the real world.

When the Israeli state comptroller made his audit and review in our offices in 1973, the subject of the "payments" came up. The state comptroller said: "Everything is true and honest, but it is forbidden to make such payments."

My reply was: "Why are you asking me? Here is the cable from our UN representative to the Foreign Office, here is the Board resolution approving the payments. If you have any objections, why don't you go to the Foreign Office to complain? After all, you and the Foreign Ministry both have the same employer. The same government that pays your salary pays theirs as well. Why bother me?"

The state comptroller responded: "You gave the instruction to execute the payment."

At the end of the audit, a month or so later, the state comptroller told me that, in spite of all the infractions and many unauthorized actions of the company, he hoped to God that there would be more Israelis like me. I considered it the greatest of compliments.

When we received the blocks offshore Gabon, our Board of Directors was very excited. Finally we had prospective acreage in a rich, oil-producing country. The directors wanted very much that the necessary seismic surveys be executed and paid for by I.N.O.C. They formed an action committee and requested a meeting with Finance Minister Pinhas Sapir. The minister hardly let them finish their plea. He said that, as far as he was concerned, Dr. Dinstein, the deputy minister, was the minister of oil. So why did they bother him (Sapir)? He had more important things to do.

The following week, the Board of Directors committee went to see Dr. Dinstein. He said, "I told you so, why did you cause yourselves to be exposed to Sapir's wrath?" "Beggars can't be choosers," Dr. Dinstein added. "Let Zvi sell part of the concession and pay for the seismic survey with this money." That was the end of the effort to do something with our own money, without demeaning ourselves by having to sell parts of concessions from the very beginning.

Chapter Thirteen

Hopes in Madagascar

In late 1969 we also applied for a concession on the island of Madagascar—the Malagasi Republic. Madagascar is the second largest island in the world, located off the east coast of Africa.

There are very large tar sand deposits in Madagascar—the Bemulenga Tar Sands. These oil-saturated sands indicate that very large quantities of oil had been created on the island, during its long geological history. This oil lost its light components when the sands containing the oil rose from the earth's bowels, the upper layers were eroded, and the oil sands became exposed to the elements. The light components evaporated and what remained is the sands, in the form of rock, saturated with tar-asphalt.

The Bemulenga Tar Sands are the second largest deposits of tar sands in the world. The largest deposits are in Northern Canada—the Athabasca Tar Sands. There is more dried-out oil in Athabasca than all the liquid oil discovered in the world during the last 150 years, including the whole of the Middle East. Unfortunately, these tar sands are hard as rock and have to be mined, crushed, heated, and the light portion, which evaporated, replaced. When hydrogen is added, the chemical reaction transforms it into liquid oil. By the end of the day, you have spent so much money on all these processes that the oil that you recover costs almost twice as much per barrel as crude liquid oil.

After more than forty years' experience with various methods of extracting this Canadian oil, the total production from Athabasca still does not exceed three hundred thousand barrels per day. Several of the major oil companies in the world, with the help of large subsidies from the Canadian government, are engaged in this effort.

When I started in the oil business in the late 1950s, I remember reading an article stating that when the price of crude oil reaches $3.50

per barrel (Middle East crude oil was then selling at $1.50 a barrel), Athabasca would become commercial. Today, when crude oil sells at $28–$30 per barrel, Athabasca is still barely commercial. It still costs twice the price of crude oil to produce, just as it did forty years ago.

Coming back to Madagascar, the existence of the Bemulenga Tar Sands indicated that large quantities of oil were present on the island during its long geological history. It was hoped that liquid crude oil could be found either further inland or at greater depths.

As there were not many applicants for licenses in Madagascar, we were allocated a very large block called Nossi Be Concession, onshore and offshore, in the northwestern part of the island.

After obtaining the license, we started looking for partners to fund and perform the seismic survey and later to pay for the drilling operations.

It was very difficult to find partners for Madagascar, as this was terra incognita as far as oil exploration was concerned. Finally we concluded an agreement with a small company named Oxoco. We had negotiated with them a year earlier, but we were looking for a stronger partner at the time. When this did not materialize we went back to Oxoco.

Oxoco had a very interesting mode of operations. Many small companies try to devise a "patent" or a niche how to enter the very costly world of oil exploration. Oxoco came up with a brilliant idea, namely to sign a full year's agreement with seismic contractors that perform seismic surveys offshore. As these companies often have slack periods between contracts, a full year's contract is very attractive to them. An undertaking to employ them for 365 days a year would rate a great discount, up to 50 percent off the regular cost. This is what Oxoco did.

After having signed such a contract, Oxoco would approach various companies which had offshore licenses and would offer a seismic survey for a "piece of the action"—*x percent* of the licensed area for the execution of *y line kilometers* of the seismic survey. The value of the seismic survey, for the license holder, was the regular cost of a survey, whereas Oxoco paid only 50 percent of such cost.

Oxoco put together a group of companies to support it. They furnished all the funds necessary for Oxoco's "patent" and received, in return, part of the rights which Oxoco had acquired in that transaction. Its main supporter was Kewanee Oil company from Bryn Mawr, Pennsylvania, our partners in Ghana. To complete the group, Oxoco would always add several more companies, in order to have the project fully funded by Oxoco's partners.

Our agreement with Oxoco in Madagascar was that Oxoco would shoot four to five hundred kilometers of seismic survey offshore Madagascar (in a seismic survey you detonate explosives to obtain the echoes from the subsurface, therefore such surveys are often described as "shooting seismic"). They would then earn 40 percent of the license area. Oxoco had three partners in this venture, who would fund the entire survey, each earning a 10 percent interest, with Oxoco having its 10 percent for free. The results of the seismic surveys were not encouraging. Oxoco and we tried, but could not find additional partners for drilling.

In 1974 there was a communist-led revolution in Madagascar, and we virtually abandoned the project.

Chapter Fourteen

North Sea Exploration and the Arab Boycott

Gas was first discovered in the North Sea in the 1960s. The first oil field—Ekofisk—in the northern part of Norwegian North Sea waters was discovered by Philips Petroleum several years later, in 1970. It proved to be a giant oil field, containing major reserves, and the rush to explore in the North Sea was on.

The area of the North Sea was divided between England and Norway by drawing a median line, on the map, roughly half way between the two countries.

The discovery of large oil reserves in a politically safe place, in Europe, considerably changed the geopolitical picture of oil supply in the world. I went to see Dr. Zevi Dinstein and told him that if Israel wanted to be involved in this safe area, we could not employ our former method of first obtaining a license, and then finding somebody willing to pay for the exploration and development.

If we were to be accepted at all as a partner in a group applying for licenses in the North Sea, we would probably have to pay an entrance fee and thereafter bear the full cost of exploration and other related expenses. I therefore suggested that he bring up the matter to the Israeli ministerial authorities and together reach a decision in principle regarding this subject. If the decision was positive, the government would then have to allocate the funds necessary for this effort. Dr. Dinstein initiated this discussion in the Ministerial Committee for Economic Affairs and shortly thereafter I received the green light to go ahead. I was promised the necessary funding.

I started to approach many companies in which we had friends and acquaintances. We talked to Monsanto, Trans Oil, Mesa Petroleum, Hamilton Brothers, and many others; each of them belonged to a different group applying for licenses in the North Sea. I received very polite replies such as, "Thank you for thinking of us," "Good to talk to you," etc., but invariably the reply was "our group is full," or a similar negative reply.

It was obvious that the companies would not have us as participants because we were a political hindrance. If any of them was not currently involved in the Arab world, one of the other partners in their North Sea group probably was doing business in the Middle East. In any case, why create a situation which might harm them in the future, especially as there were many more "buyers than sellers." Many more companies wanted to explore in the North Sea than the number of blocks being allocated.

I also went to see the British Government's department in charge of oil exploration. They were very polite, and said that if we were to present our financial competence and submit an application, they would consider it. My feeling after the meeting was that the chances of being accepted by them were close to zero.

The only firm that was ready to talk to us was a very small British company, Ball & Collins, to whom every partner was very important. Keith Collins, its founder, major shareholder and president, was a New Zealander.

Collins had served in the Middle East during World War II and had reached the rank of a major. After the war, he married an Egyptian girl from a very prominent family and remained in Egypt. He became a supplier of food to the British forces in the Suez Canal area and made a small fortune. When President Nasser came to power, he confiscated the business. Collins moved to Libya, where he became a food supplier to the companies drilling for oil in the Libyan Desert. Keith made another fortune there. When Muammar al-Gadafi came to power in Libya, Collins had to leave again. However, in Libya he became enamored with the oil exploration business and, when he arrived in England, he created a small oil company and called it Ball & Collins.

Collins had an interesting partner, an American by the name of Roland Shaw. Roland was huge, almost seven feet tall, with a weight to match. He had been a pilot in the American Air Force during World War II and remained in Europe after the war. For several years he worked in Paris for D.D. Feldman who, in the 1950s, was a famous oilman. It was rumored that Shaw was also working for the CIA, a rumor which he never confirmed or denied.

When I first met Roland, he told me that he did not like Israelis, or rather, that in his view we did not belong in the Middle East. He said that the Jews, after being absent from Palestine for such a long time, did not have rights there. In his view, Palestine belonged to the Arabs. This did not prevent us becoming good friends on a personal level. He told me one day that the only Israeli that he had met before me was a man called Adin Teilhaber—at the London School of Economics. That was a most unusual coincidence. Adin was my wife Rachel's serious boyfriend for some time during World War II, when we both served in the British Army. Adin was a champion runner and an excellent sportsman. He was sent by the British to officers' school and later served as an officer in the Jewish Brigade. The fact that the only Israeli Roland knew and liked was my wife's former boyfriend was really odd.

Roland remained in Europe after the war and joined Keith Collins in Ball & Collins Oil Company in London. After Keith sold his interest in 1974, Roland succeeded in building Ball & Collins, which changed its name to Premier Oil, into a substantial oil company with world-wide interests.

Premier Oil was an empty shell listed on the London Stock Exchange, and Brian Russell, of Charterhouse, who, raising money for both of us, Ball & Collins and I.N.O.C., offered us both to buy a listed shell registered on the London Stock Exchange. The plan was to inject our assets into such a listed company. This procedure would save preparing a prospectus and going through the lengthy and costly efforts of registering a new company and getting Stock Exchange approval. It is an efficient way of obtaining a listing for a new company and raising money for an already listed vehicle.

Charterhouse was successful in raising money for Ball & Collins, and therefore they ended up owning Premier. Roland Shaw succeeded in making Premier into a successful oil company with substantial North Sea oil and gas production. For these services he was awarded a CBE (Commander of the British Empire), one of the very few Americans with this British title.

* * *

This period of late 1970 and early 1971 was the very beginning of the oil rush into the North Sea. Each foreign group was looking for a British "flavor," in order to be favorably considered by the British Government. Suddenly Ball & Collins became a desirable commodity.

At the end of 1969, in one of the first rounds of block allocation in the North Sea, Ball & Collins received six offshore blocks. Unfortunately, the company did not have any geological know-how or funds to acquire such geological information and advice. As a result they chose blocks that proved later to be barren.

* * *

Another small company, Ranger Oil of Calgary, Canada, was allocated six blocks in the same round. This was in 1969, about a year before the major Ekofisk oil field discovery. The British Government allocated blocks to small companies in that round. Ranger Oil, which had a good geological staff, chose its blocks very wisely and discovered one of the largest oil fields in the North Sea, the multi-billion-barrel Ninian Field. Ranger Oil thus became a successful and rich oil company.

Ranger Oil was headed by a man called Jack Pierce, whom I had first met in Tel Aviv in 1967. He came together with Charles Bronfman (one of the wealthiest Canadian businessmen) to advise the Israeli Government on how to deal with the Sinai oil fields it had seized in the Six-Day-War. I saw Jack again in 1969 when he came to my hotel in New York to discuss the possibility of Ranger's joining us in Africa. After lengthy discussions, Jack decided to stay in the "civilized world," first Canada and then the North Sea. He obviously made the right decision. After he died in the early 1990s, an oil field newly discovered by Ranger Oil in the North Sea, was named after him: the Pierce Oil field.

* * *

Back to Keith Collins. By the time I approached him, he already had his first six blocks for which he found American partners. He was getting ready to submit applications for additional blocks in the new round in 1971. As his firm was very small, any new partner was important to him.

I also contacted Sir Charles Clore, a leader of the British Jewish community and one of its most important businessmen. He was also the owner of the Selfridge's Department Store. I suggested to him to join us in any North Sea group that we would eventually end up with. He agreed.

We reached an agreement with Keith Collins to join his group and signed a contract to that effect. A few weeks later, Collins called me urgently in Tel Aviv and said that we had to meet urgently in London. When we met he said: "Zvi, listen to what has happened and tell me what to do. A German company which owns a refinery that processes Libyan oil was willing to join our group. They are very important to me as they are adding a lot of strength to our group. When they heard about you (I.N.O.C.) they said it's either us or I.N.O.C. They said that they would jeopardize their relations with Libya by being partners with the Israeli National Oil Company. What shall I do?" he asked. Obviously, we had no choice but to give up our participation and leave the group. I had the unpleasant task of informing Sir Charles Clore of our predicament but promised to include him if we were to join another group.

The British Government had various criteria for approving a group of applicants for North Sea exploration. As is customary in Britain, such rules are neither written nor spelled out. They are a result of various considerations and very long experience. In the case of the North Sea, the criteria were the financial strength of the partners, their know-how and experience, a British involvement, the participation of a drilling contractor in the group (there was a great shortage of offshore drilling equipment at the time), and the experience of the group in offshore exploration. All these considerations were factors in the decision-making process, and in the allocation of the most promising areas.

In this regard of British customs, I heard an interesting story years later. Adrian Evans, one of the directors of the bank that years later assumed ownership of our company, and who now is the deputy head of Lazard Frères in London, told me about an interesting episode. When Adrian came back from the US in 1972, after having been a vice-president in Citibank in New York, he was offered a very good job in London with an excellent salary. He consulted a friend of his in the Bank of England: "I was offered this job. What do you think?" His friend did not say a word but made a certain movement with his nose. This was a sufficient response for Adrian to refuse the offer. Adrian told me that, had he ignored that hint from an official in the Bank of England, it would have been the end of his career in the City of London.

Chapter Fifteen

Diamonds in South Africa

As a result of our relations with Keith Collins, we made the acquaintance of an investment house called Charterhouse Japhet. They were then working to raise some money for Ball & Collins. After we met, they offered to perform the same service for us.

Their representative was a very tall lawyer, by the name of Brian Russell. He had once worked at Shell and was familiar with the oil business. He was very attracted to the prospects of our concessions in Africa. After hiring an international firm to appraise the value of our licenses and receiving their report, he became very optimistic that he could raise fourteen million dollars for us.

One of Brian Russell's misgivings was that he might be faced with some difficulties talking to the investment community. They might be concerned that all our licenses were in black African countries. He therefore expressed the opinion that it would be very beneficial for us to dilute our black African involvement with another involvement in a "whiter" area.

At about the same time John Christensen, our geologist from Rome, came to us with a proposition regarding a diamond exploration venture in South Africa. There was a large farm in South Africa where diamonds had been mined and produced in the 1930s. He showed us South African government records to that effect and said that there were good prospects for finding more diamonds in that area. The rights belonged to two South African geologists who were willing to sell us an option for one third of their right for $130,000. We had an additional option of purchasing another third at a much higher price. This second option would obviously only be exercised if the results of the initial surveys and exploration were favorable. We had one final option, to acquire up to 90 percent of the venture, for many millions of dollars.

We had considered the possibility of diamond exploration in the past. Mineral exploration was in our original mandate given by the ministerial committee. Another contributing factor was the existence of a very large diamond cutting industry in Israel, the second largest in the world, that would have provided us with a ready market, development capital, etc.

John Christensen's proposal was very timely and fell on receptive ears. I presented the idea to Brian Russell, he was enthusiastic. London is the world's marketing capital for diamonds. In Russell's opinion, it would therefore win over his prospective investors and add a great deal to our package. The name of John Christensen, who recommended this venture, added a lot of credibility and seriousness to the project. He was well known in the mining industry as the discoverer of very large deposits of diamonds offshore South West Africa, now Namibia.

Although $130,000 was not a very large sum, we did not have it. I met with Dr. Dinstein, and described the diamonds opportunity to him. Hanan Yavor, one of our directors and my dear friend and supporter, participated in the meeting as usual. Dr. Dinstein, realizing the importance of the venture, promised us the necessary sum to buy the first third of the diamond option. This was the first time he allocated money to our company.

Several years later, when we were already operating from England, we had to let go one of our senior geologists who then returned to Tel Aviv. In Israel he found a "friendly" journalist who published a series of articles blaming the company, and me for getting involved in a "diamond folly." The same geologist had been the strongest advocate of our involvement in this diamond venture. He was in charge of the fieldwork in South Africa, was excited about the prospect, and constantly submitted glowing reports. But such favorable information does not make interesting news. The only news worth publishing is a claim that you did something wrong. The truth be damned.

We did find some diamonds in South Africa, but the project did not show signs of becoming a substantial money earner. When the Charterhouse affair fell through, as will be described later, we stopped the fieldwork in South Africa but still retained our legal rights.

It is important to note that two separate, independent appraisals of our assets, years later, gave us a full credit for all our investment in the project and listed it among our assets, at full value.

Chapter Sixteen

Signal Oil & Gas and Charterhouse's Treachery

Signal Oil and Gas was a large independent oil company established in California at the beginning of the twentieth century. It found a very large oil field on a hill near Los Angeles, called Signal Hill, and the company adopted the name of Signal Oil and Gas.

Signal was one of the first American independent oil companies to "go foreign," i.e., to explore outside the borders of the United States. They joined the Iranian consortium in the mid-50s and went on to discover and produce oil in Kuwait. Later they were involved in Argentina and other places in the world, including the southern part of the North Sea, where they found gas in the mid-1960s.

Signal was our neighbor in the offshore of Ghana. They were the only people to discover oil in Ghana, a very rich well which produced ten thousand barrels a day, in an area called the Salt Ponds. Unfortunately, six other offshore wells they drilled in surrounding areas were dry, and Signal abandoned the whole project. To this day Salt Ponds is the only producing area in Ghana, although it is now very small and the oil of poor quality. It is being worked by the Ghanaian National Oil Company.

For many years I did not know that Signal had a long (almost fifteen years) secret connection with the State of Israel. I discovered this only in 1971. As they were our neighbors in Ghana, and as we also had other mutual interests in the state of Gabon, we sometimes visited their offices in London, to discuss various projects. One day, in the winter of 1971, after the Ball & Collins North Sea debacle, I visited Signal's offices in London together with Dr. Avihu Ginsburg, the head of the Israeli Geophysical Institute, who was one of our directors. We started chatting about various countries and at some point I asked them about their future plans in the North Sea. They were the pioneers of North Sea exploration from the early 1960s and had had an active exploration office in London for the past ten years.

Their reply to my question regarding their North Sea group was as expected, i.e., the group was complete and was not looking for additional partners. During the conversation the name of Dutch Lortcher, their chief executive and chairman, came up. According to his people in London, he was the all-powerful "dictator" of the company. The name rang a distant bell in my mind. I somehow thought I might have heard this name mentioned in Israel before.

When I arrived back in Tel Aviv, I asked Dr. Dinstein whether he knew Dutch Lortcher. "Oh," he said, "he is a very good friend of mine." I suggested to Dinstein that we write a letter to Lortcher describing I.N.O.C.'s international activities, and voice our desire to enter North Sea exploration. I added that, as I.N.O.C.'s licenses adjoined Signal's holdings in Africa, perhaps Signal could also consider an investment in I.N.O.C.

A few weeks later, Lortcher's reply arrived. He did not exactly understand what we wanted from him, but as he was soon coming for a short visit to Israel, he suggested we discuss it face to face. I met Dutch Lortcher when he arrived in Israel. There was a positive chemistry between us and we chatted for a long time. The idea of investing in I.N.O.C. was as alien to him as to all the other oilmen with whom I had discussed this idea. Oil companies do not, as a rule, invest in the shares of other oil companies—primarily for tax reasons. Whereas their own dry holes can be written off immediately as an expense in their profit and loss statement, they are not allowed to write off losses resulting from dry holes of companies in which they own shares.

As to participation in the North Sea, there was a remote possibility that Lortcher could include us in his group, and give us a small participating interest. He listed several preconditions we would have to observe. First and foremost was complete secrecy. The Signal consortium included an offshore drilling company called Santa Fe which was very heavily involved in the Arab world. Santa Fe was essential to Signal as the mandatory drilling contractor vis-à-vis the British Government.

Secondly, we must find ourselves a British cover. "This way," he said, "we will kill two birds with one stone. On the one hand, you will be completely hidden, and on the other, the group will have a British participation. Such participation will help us to get the blocks we desire."

As this was the period of our "love affair" with Charterhouse Japhet, I asked them to provide us with the necessary cover by having

Charterhouse as a consortium member, holding our interest in trust. I went to see Sir Charles Clore and I concluded with him that he would receive a 2 percent participation, while we would have 7 percent. We left 1 percent for Charterhouse, which we were sure they would demand for the cover.

In May 1971 I received a telex from Charterhouse confirming their agreement to have their name submitted to the British Government as a 10 percent holder and confirming that they would hold our interest in trust. This telex was immediately forwarded to Signal.

On the 1st of July, my wife's vacation from school started and we went together to London. When we turned on the television, there was a news item that the chairman of Charterhouse had died of a heart attack while playing golf. Rachel had a premonition that his demise would affect us very badly. I said: "What are you talking about? How could this disrupt my negotiations with Charterhouse?" "Believe me," she replied, "my woman's intuition tells me this is very, very bad."

A few days later I visited the Charterhouse offices to meet with Malcolm Wells, one of their directors. During the conversation, Wells told me: "Mr. Alexander, we have a problem. We couldn't care less if everybody outside the Signal North Sea Consortium believes that we own the whole 10 percent interest, but we should feel bad if any of our colleagues within the consortium are kept in the dark. We would really feel much better if the members of the group knew the truth."

I said to him: "Dear Mr. Wells, it is very late in the game to advise me of your feelings. Signal has already informed the British Government that you are a 10 percent member of the consortium. You cannot, at this late stage, change the rules of the game. You also know that the only reason we have asked you to front for us is that one important member of the consortium cannot be politically associated with us."

He responded: "I understand you but you must understand our problem as well. By the way, one of your countrymen is visiting us now and we are having lunch with him." This was Yaacov Meridor, the head of Maritime Fruit Carriers, the first Israeli company to raise public money on Wall Street. My general feeling was that there were difficulties with Charterhouse, but that they could be solved.

Several days later, Brian Russell came to our flat in London. The flat served primarily as a mail drop and as a meeting place for people and companies that did not want it to be known they were associated with an Israeli company. When my wife walked in to be introduced to Brian, he threw his briefcase on the table and said: "They dropped your project and I have resigned from Charterhouse."

He handed me a letter from Charterhouse stating, in general terms, the following: "We have checked our situation, we won't be able to raise fourteen million dollars for you. It is too large a sum for us. Consequently, we won't be able to provide a cover for you and represent you in the North Sea Consortium. We suggest that the same party that will raise the money for you will also provide the cover and represent you in the North Sea. Thank you very much, etc., etc. Let's meet again in other circumstances. Good-bye."

Suddenly we had lost our cover, and if I did not find another immediately, we would lose our only opportunity to enter the North Sea exploration. I asked for a prompt meeting with Sir Charles Clore and told him that he must furnish me with a British company. If his lawyer would provide me immediately with the name of a financially strong British company, I would try to substitute it for Charterhouse. Maybe it would work. Obviously the first demand of his lawyers was to increase Clore's participation to 4 percent, up from the 2 percent. I agreed. This left us with 5 percent. But I needed the British company as soon as possible.

Clore's chief lawyer and consultant, Leonard Sainer, who was also one of the heads of the Jewish community in London, dragged his feet. It took him five days to give me the name of the new company. I felt as if my blood was draining from my body. I did not inform either Israel or Signal of my predicament. I suffered all this anguish by myself. On the sixth day, Sainer informed me that they had a company, Princess Investments, with a balance sheet of fifty million pounds, including large real estate holdings. Princess Investments was known as a reputable and substantial company. Sainer asked: "Who will give us a guarantee that you will pay your share of the exploration costs?"

"Is the Government of Israel not a sufficient guarantee for you?" I countered.

"No."

"Then whom do you want?"

"At least Bank Leumi." (the largest bank in Israel)

I said: "You will get it."

I placed a phone call to Roland Schwab, the head of Signal's London offices. I wanted to notify this change of the participating company to a lower-level official in Signal's hierarchy, in the hope that it would cause less of an uproar. I found out that Schwab was in Los Angeles, at Signal's headquarters, and caught him there. I said to him: "Roland, these Charterhouse people are really no good. There are always new problems with them, so I decided to get rid of them and replace them with a different company that will be easier to work with. I found a very good British company, which has assets in excess of fifty million pounds. Its name is 'Princess Investments,' and it would be a pleasure to work with them. May I suggest that you delete the name Charterhouse, wherever it is listed, and insert 'Princess Investments' instead." Roland Schwab agreed. I felt that I was saved.

I told my wife that I needed some fresh air. I was going for a walk in Hyde Park and would be back in an hour. When I returned, before eleven o'clock at night, Rachel asked me: "Where were you?" I asked what had happened. She said: "Dutch Lortcher called twice from the United Sates and Zevi Dinstein twice from Israel. Both of them asked that you return their calls immediately."

I phoned Dutch Lortcher. He said: "Listen, we told the British Government that Charterhouse is our partner; so it is either Charterhouse or you are out. We cannot at this stage change names just because you want to." I started to make explanations, but he did not want to listen. Although it was already one o'clock in the morning in Israel, I called Dinstein but did not find him. At 2 a.m. London time, I called Lortcher again in Los Angeles. I said: "Dutch, I never lie, not because I am so honest, but because I do not want to have to remember what I said last time. If you tell the truth, you do not have to remember what you said previously. Now, please listen to me and let me tell you the whole story." When I finished, I felt that the mood had changed 180 degrees. Once I had told him everything, I sensed that a feeling of fairness took over. By the way, I have almost always felt that fairness was a very important trait of the American psyche, especially of those Americans I met in the oil business.

Dutch said: "Look, it is very difficult for us to make the change now. Tell Charterhouse that the contact you created for them with Signal Oil maybe be worth millions of dollars to them. We are always looking for finance for large projects. They are an investment house which wants to grow, and by becoming one of our bankers they will benefit considerably. The problems are not so difficult; try to influence and convince them, and let us see what happens next." I said: "Fine. I will call them first thing tomorrow morning."

I phoned Dr. Dinstein the next morning and he was furious. He said: "Do you think that you are a contractor for successes only? How dare you keep such a burden to yourself for a whole week without notifying anybody?"

"What would you have done?" I replied. "What would you have been able to do? Would you have told me that Clore would find me a company, or Sir Isaac Wolfson would find me a company? I found a company. How could you have helped me if I had told you? Would you also have been mad? Would this have helped? I had to find a solution and I did."

Dinstein was not mollified. "I am asking you, and I am demanding from you, that you won't carry such burdens alone. A man needs to consult, needs help and advice. How could you do such a thing?"

In any event, by now both sides, the Israeli Government and Signal, were in the know, and I did not feel like I was stranded in the desert, all by myself.

I called Charterhouse and spoke to a director of the bank, by the name of Dalton, who was Brian Russell's superior. I began: "Mr. Dalton, I would like to see the head of the bank this morning."

He said: "The head of the bank has died."

"Mr. Dalton, I do not plan to go to the cemetery. Somebody must have been appointed as the new head of the bank."

I met with Dalton the same day and told him that I had talked at length with the chairman of Signal. The latter had told me that Charterhouse would benefit immensely from the association with Signal if they kept their word and their undertakings. On the other hand, if they refused to go forward, they would have to remember that we had their written undertaking promising to front for us. Due to their refusal to go forward, I.N.O.C. would lose its participation in the North Sea. They might be liable to a law suit and a demand for substantial damages. As to Signal, they really did not think that Charterhouse could back out now. I must have talked with Dalton five times that day. They convened a special Board meeting at Charterhouse to deal with this subject. After that meeting, Dalton informed me that they would be happy to front for us, subject to the other members of the consortium being notified—the same answer that I had heard previously from Malcolm Wells. I reported to Lortcher after each conversation with Dalton. I had a feeling that after each conversation Lortcher was becoming a better and closer friend of mine.

After a long week of further conversations and discussions with Charterhouse, Lortcher said: "Let us leave these S.O.B.'s with the 1 percent interest that you gave them, as their name appears in the applications to the British Government. We shall try to include your Princess Investment for a 9 percent interest. We shall make this change as close as possible to the date of the submission of the application. We shall try and do it only forty-eight hours before the deadline, so that Santa Fe won't have enough time to investigate what has caused the change. The last forty-eight hours are usually devoted to the final discussions among the partners in the group about the desirability of various blocks and other administrative arrangements. We shall do it in mid-August when these activities will be at their peak, the turmoil will be the greatest, and let us hope it will work." I said: "Wonderful!"

One has to remember that, at that stage, we were only talking about the composition of the consortium and its application for a number of specific blocks in the North Sea. There was obviously no certainty that the consortium would be allocated the particular blocks that they desired.

One day in mid-August the phone rang in our home in Zahala; Dutch Lortcher was on the line. He said: "Zvi, the ploy did not succeed. Yesterday afternoon, thirty hours before the application had to be submitted to the Government, we listed Princess Investment. This morning, Santa Fe representatives came in and informed Peter Rainer, Signal's vice-president for exploration, that they had investigated and discovered that Princess Investment belongs to a Jewish gentleman, a great supporter of Zionism and of the State of Israel. Santa Fe does not agree to be a member in a consortium with a company connected in any shape or form with the State of Israel."

I said: "What do we do now?"

He replied: "This is what I am going to do: Clore must be out completely. We shall leave Charterhouse with 1 percent. The remaining 9 percent we shall divide equally between the four members of the consortium. Signal will therefore receive an additional 2.25 percent. I will reduce Signal's original interest by 2.75 percent, which will create a total of 5 percent. (2.25 + 2.75). I will create a new company, to be named North Sea Petroleum, which will hold this 5 percent interest. I will explain that I am dividing our interest into two vehicles, for tax considerations. These 5 percent of North Sea Petroleum I will keep in trust for you and down the road I hope we will be able to transfer it to you officially."

I was bowled over: "Dutch, thank you very, very much."

We were immediately swamped by a large number of angry cables and messages from Sir Charles Clore. He was very much offended, and justly so. It is very possible that if Clore had gone with a group other than I.N.O.C., one that did not include a similar "Santa Fe," he could have been involved in the North Sea exploration without difficulties.

This debacle brought Dutch Lortcher and me very close. I had opportunities during our many conversations to describe all the problems and difficulties I encountered in raising money. These difficulties were primarily based on the fact that we did not have an oil company shareholder. I think I probably convinced him to consider becoming our savior.

* * *

Our effort in 1969–1970 with Planet Oil was of a completely different nature. Becker, the investor, might have considered Planet to be an oil company but in truth Planet was a midget, with no money, experience or technical staff, etc. Signal Oil, on the other hand, was one of the largest and most reputable independent oil companies in America, known throughout the world. If Signal, a billion dollar oil company, would invest in our company, it would immediately give us enough credibility and standing on Wall Street.

I had a feeling that my message was starting to come through, and that I had found a sympathetic ear. Maybe Dutch did not understand why I needed it so much, but he was ready to consider helping us. In August, I met with Dutch in London. I convinced him to check out our holdings to see whether they were of interest to Signal. Dutch called his vice-president for exploration, Peter Rainer, and told him to go to Israel and take along with him as many people as he needed. Dutch said: "Check Zvi's company from top to bottom and from left to right. I do not want to hear from you reasons why I should not get involved. What I do want is to obtain facts. The decision I will make."

At the end of August, five Signal representatives, including Peter Rainier (the vice-president), Roland Schwab (the head of the London office), one more geologist, one geophysicist and one lawyer arrived in Israel. They checked every map we had, every seismic section, every concession agreement, every contract and all other relevant documents. They stayed in the country for ten days.

They reported to Lortcher that, from the point of view of exploration, licenses and prospects, the picture was very positive. The political implications were negative. Peter Rainier was dead set against their involvement. His main concern was that this "contamination" might jeopardize their future.

In spite of Rainier's opposition, Dutch instructed him to send a lawyer and a geologist to check our licenses in Africa, to ascertain that the concessions were in good standing, and to ensure that work commitments were executed on schedule.

The next step was to establish the value of our holdings. We invited DeGoyler and McNaughton, the most prestigious valuer of oil assets, to assess the market value of our holdings. Jim Glover, one of DeGoyler's senior geologists, spent over two weeks in our offices . He came up with a figure of fifteen million dollars. His report was forwarded to Signal. That figure of fifteen million dollars later formed the basis for our future money-raising efforts.

It must be realized, that from a purely business point of view, Signal's hoped-for investment of several million dollars was well outside their usual business practice. There was no reason why such a big company would invest in an insignificant company which, furthermore, might cause it political damage. It was achieved thanks to Lortcher's personal involvement and his willingness to help me personally, despite all his difficulties.

* * *

It is important at this stage to explain Signal's secret involvement with the State of Israel in which Dutch Lortcher played an important role. In 1954, after the fall of Mossadeq in Iran, and the return of the Shah, which was partly engineered by the CIA, American oil companies for the first time became part of the Iranian consortium. This consortium was formerly completely controlled by the British. In addition to the American major oil companies which joined the Iranian consortium, a group of independent oil companies was formed, called the Iricon Group, which took an 8 percent interest in the Iranian Consortium. Signal Oil was part of the Iricon Group. As none of the members of the Iricon Group had a market for their share of the Iranian oil, Signal's executive vice-president, J. Howard Marshall (our partner in Nigeria in the late 1960s, see chapter 5), came up with a brilliant idea, namely to sell Signal's Iranian production secretly to Israel. Israel was very hungry for Middle Eastern oil, which they could not purchase for political reasons. Following the supply contract of Iranian oil to Israel, Signal bought 33 percent of Paz Oil, the largest oil marketing company in Israel.

Both of these transactions were state secrets in Israel, known only to a handful of people. The ownership of Paz was thought to be in the hands of Max Fisher, a very wealthy Jewish businessman from Detroit and a great supporter of Israel, who was representing Signal. Actually, Fisher did not own one single share of Paz, but he acted as if he owned it all (see chapter 4). The other reason that this Israeli connection had to be kept in complete secrecy was that Signal was at the same time producing oil in Kuwait. Probably fewer than five people in the whole Signal empire knew about this duality.

Signal was a very profitable company. With the profits of their oil business in the 1960s, they acquired two very large industrial enterprises—Mack Trucks and an aerospace company called Garret Engineering. They also purchased large real estate holdings in Hawaii. Obviously, to acquire all these assets, they had to borrow a lot of money. The fall of the stock market in the early 1970s caused them to change their course of action. They decided to sell their refining and marketing interests both in the US and in Europe and to concentrate on oil exploration and production. As a consequence, their ownership of Paz became unimportant.

Dutch Lortcher and Signal's vice-president of legal affairs, Bill Miller, arrived in Israel to negotiate the sale of their interest in Paz. The negotiations with the Government took longer than expected and I had an opportunity to spend a lot of time with both Dutch and Bill Miller. By that time Dutch had received the report of his delegation to Israel. He knew that we had valuable concessions. We took them both on various trips to show them the country, and they also visited me at home and in the office several times. Surprisingly, they were late for their first visit to our offices. Usually, Dutch was very punctual. Eventually, I found out what had happened. I had told him that our offices were on Petah Tikvah Road, which is one of the main avenues in Tel Aviv. He forgot to mention "road" to the taxi driver, who took him to Petah Tikvah, a town ten miles east of Tel Aviv.

Throughout their visit I tried to convince them to consider favorably an investment in our company. During the first few days, Dutch maintained that Signal could not consider any investment in the shares of the company. They would, on the other hand, be ready to join us and be our partners in Gabon, as well as possibly in Ethiopia and other places.

I also asked for help from a good friend of mine, Joe Boxenbaum, who was also a good friend of Dutch Lortcher. Joe was a member of the Paz Board of Directors, representing Signal's shareholding. Joe was a very wealthy industrialist in Israel and the representative of Chrysler motor cars in the country. He was an American who came to Israel after the Second World War. He served in the Israeli Army as a high-ranking officer, and remained in Israel thereafter.

On the eighth day of their stay in Israel, Joe told me that Lortcher had said that he was going to make me very happy. Later that day Lortcher contacted me: "Listen, it is not 'yes' as yet, but I am ready to consider the investment. I have to go back to Los Angeles and speak to other Board members, as this is a matter for the Board to decide. Let us see what I can accomplish. It will not be easy, for sure."

During the whole month of October, we were in a state of great apprehension, hoping against hope that Lortcher would come back with a positive answer.

At the end of October, he phoned me from Los Angeles and said: "Bring your lawyers, we can start writing the agreement." I was in seventh heaven.

Adi Ephrat, our lawyer, and I arrived in New York. We met Dutch, Bill Thompson, Signal's executive vice-president, and Peter Rainier at the Dorset Hotel, where Signal had its permanent suite in New York. Even at that late stage, Peter Rainier called Dutch to the next room to try to convince him not to go ahead with the deal. Luckily, Dutch Lortcher was a man who made up his own mind, and once he made a decision, he stuck by it.

In the last week of November, 1971, Adi Ephrat and I were invited to Los Angeles to conclude the agreement. This was the week of the Thanksgiving holiday in America. We were working in Signal's completely empty offices. Thanksgiving is celebrated on the last Thursday of November, but because of its proximity to Saturday and Sunday most people take a four-day vacation, including Friday. Adi and I decided to catch some fresh air in the streets of Los Angeles. Signal was located near an office of Merrill Lynch, the largest brokerage house in the US. We walked into their office to watch the tape on which the transactions of the New York Stock Exchange were recorded. The Stock Exchange was open on the Friday of Thanksgiving weekend. We looked at the tape and were amazed to see only plus signs.

The market had been terrible for many months before Thanksgiving, yet suddenly, on that day, when most people were on vacation, the market reversed its direction. I will never forget this experience, which was another example of how unpredictable the stock market is. It reminded me of a book I had read many years earlier about Paul Getty, the wealthiest oilman of all times. He was asked for his views on the stock market, and he replied that the difference between a speculator and an investor was that the speculator bets on the weather (which obviously is completely unpredictable), whereas the investor is betting on the climate, knowing for sure that after winter comes the spring. If you invest in a good company, with good management, you will over time end up in the 'plus' column.

Although Dutch was planning to invest three million dollars for a 20 percent interest in our joint company, the Board of Signal decided to limit the investment to one and a half million dollars for a 10 percent interest. The agreement was based on the formation of a new company to be registered in Bermuda, for tax reasons, and to be named PEDCO—Petroleum Exploration and Development Company, in which Signal would hold a 10 percent interest and I.N.O.C. 90 percent. There were many other clauses in the agreement. It included Signal's technical assistance to PEDCO, options for Signal to participate in PEDCO's future concessions (which obviously was to our advantage as well, having Signal as built-in partner), and other provisions. Dutch Lortcher was to become the chairman of the new company and I the managing director.

The most important condition was that we were obligated to raise additional finance of at least five million dollars within six to nine months, in order that the company would be adequately financed for the immediate future.

Ken Bialkin, our lawyer, introduced us to CBWL Hayden Stone, an important and prestigious investment banking firm in New York City. They undertook to raise the five million dollars for us. It was to be a private placement and not a public issue. Two of their representatives, an elderly lawyer, Orvis Sowerwine, and a younger banker, Jack Byrnes, arrived in our Tel Aviv office to prepare the placing documents. Sowerwine had an interesting personal history. He was a successful Wall Street lawyer during the 1950s and 1960s. A large part of his work was for Hayden Stone, an almost one-hundred-years-old prestigious Wall Street firm. He was invited to leave his law practice and become a partner in Hayden Stone. As a partner, he had to contribute a share amounting to half a million dollars, which in today's money is more than five million.

He invested all his savings and borrowed money from his family to come up with this sum.

In 1970, Hayden Stone decided to install a new computer system to do all the back room work. This work had previously been done by hand, by a large number of employees. Computers in 1970 were considered to be a very modern and advanced tool. It did not occur to Hayden Stone that initially they would have to run both systems, the manual and the computer, concurrently, until they had learned how to use the computer correctly and efficiently. When they installed the computer, they immediately eliminated the manual work. Within two months, Hayden Stone was close to declaring bankruptcy. Their accounts were in a mess, because of the inexpert use of the computer. Clients who owed them money were not asked to pay, whereas Hayden Stone had to pay its creditors. After one hundred years, the old company was about to go under.

Orvis Sowerwine lost all his investment. In desperation, he planned to become a fisherman for a living. He and his family lived in New Jersey near a lake, and he thought he could support his family catching fish.

Ken Bialkin, our lawyer in New York, was instrumental in saving Hayden Stone. He knew four energetic young men who had established a very small banking firm called CBWL, named for its four founders: Cogan, Berlin, Weill and Levitt. He introduced them to the Hayden Stone situation and they took over the proud firm of Hayden Stone for free. They renamed it CBWL Hayden Stone. Obviously, they took complete control. By the way, after some five or six mergers and acquisitions, CBWL Hayden Stone is today the mammoth firm of Salomon Smith Barney, a part of Citigroup, whose head is Sandy Weill, the "W" in CBWL. The other members of the CBWL Group have done very well too. Levitt was the president of the American Stock Exchange, Cogan is one of the owners of Sotheby's.

* * *

Both Orvis Sowerwine and Jack Byrnes liked what they saw in our offices in Tel Aviv, and a month later we started the "road show" in the US, to meet the prospective investors.

We had a very impressive group, namely two representatives of CBWL Hayden Stone, Jim Glover, the geologist from De Goyler and McNaughton, who appraised our licenses at the value of fifteen million dollars—hence the one and a half million dollars of Signal for a 10 percent interest. De Goyler and McNaughton were, and still are, the most prestigious and reliable appraisers of oil properties in the world. Their reports are accepted verbatim by the banks when providing credit to oil companies. Therefore their representative's presence at meeting with prospective investors was very important. I asked Dutch Lortcher to have Bill Thompson, the executive vice-president of Signal, the number two man after Lortcher, join our group. Reluctantly, Dutch agreed and Bill Thompson joined us for the two weeks of the road show.

The stock market turned bad again. The people at Hayden Stone had either not done their homework correctly to identify suitable investors, or they no longer had the right connections and credibility with serious investors. We travelled for two weeks from place to place but had little to show for it. A small success is meaningless—either you raise the whole amount you specified or you don't. It is useless to raise two million out of four because those who committed the two will refuse to pay up. It is an all or nothing game.

I was terribly disappointed. I had hoped that Signal Oil's name would be a great attraction to investors and my troubles would be over. But the proud name of Signal had not helped enough and we were still in the middle of nowhere. We might even lose the Signal commitment.

With the meteoric rise of the stock market during the whole of the 1990s, the young people working on Wall Street did not comprehend the meaning of a falling market and its effect on investors. In a slump you cannot sell gold bricks even for the value of their gold. This reminds me of the dictum of Warren Buffet, the greatest investor of this century, that an investor must remember that his partner is Mr. Market. Sometimes Mr. Market is in euphoria and sometimes he suffers from depression. In my case, Mr. Market was almost always in a depression.

Two years have passed since I wrote the above observation about the market. What a change! The young dot.com entrepreneurs, who were flying high when the sky was the limit, now see what a collapsing market means. Billions have disappeared into thin air and nobody sees the light at the end of the tunnel yet.

* * *

In May 1972, Signal had a "closing" in New Orleans with John Mecom's company. "Closing" is the American method of concluding a contract, where all the relevant documents are presented and approved at one session. Signal bought 51 percent of Mecom's properties and this removed Mecom out of bankruptcy.

If there was anyone who deserved the attribute "larger than life," John Mecom was such a man. He was one of the wealthiest people in Texas and, at one stage in the late 1960s or early 1970s, he virtually tried to buy control of the whole State of Texas. Mecom bought hotels, a Texas football team, the leading newspaper, and so on and so forth. He obviously borrowed a lot of money from the banks for this buying spree. Finally, the old establishment of Texas, especially Houston, which was unhappy with and apprehensive of Mecom's actions, decided to pull the rug out from under him. Using their influence on the Texas banks, they called in his loans and forced him to declare bankruptcy. His lifestyle was such that the bankruptcy court allowed him to draw five hundred thousand dollars a year for living expenses, which in today's money is worth more than five million dollars.

Mecom's fortune was made when he discovered a five hundred million barrel oil field in 1944, near Lake Charles in Louisiana. There was a belief among oilmen that oil can be found only on the northern side of a salt dome. Mecom drilled on the "wrong" side of a salt dome, the southern side, and discovered a tremendous oil field. One of the first things he did when World War II ended in 1945 was to go to Europe and buy a shipload full of antiques. His house, his hotels, and his children's houses were filled with antiques. Still, he ran short of space and had to house the rest of his treasures in two giant hangars of blimps (large balloons), which he rented in Houston.

After Mecom concluded his agreements with Signal, he and I met often and we became good friends. He kept urging me to arrange for an Israeli commando unit to cross the border from Eilat, on the Red Sea, to Aqaba, in Jordan, in order to retrieve his drilling rigs which had been confiscated by King Hussein. He said: "It is not a problem for you to do, and we will divide the equipment between us."

John Mecom was the first independent oilman to "go foreign" which, in American oilman's language, meant exploring for oil outside the US. This was quite a novel idea in 1950s America. Only the major oil companies, but no independent, had ever dared to explore abroad. Mecom went to Yemen, Jordan and other places in the Middle East to drill for oil. He drilled two deep holes in Jordan, in Ramallah and Halhul, then in Jordan and now in the West Bank. Both wells unfortunately were dry holes. He came to the conclusion that there is nothing more of interest to explore in Jordan. But the Jordanian government demanded that he fulfill his obligation to drill more holes. When he refused, they confiscated his equipment. Hence his suggestion that we retrieve his rigs.

There was nothing that Mecom enjoyed more than a bargain. One Saturday we had lunch with Dutch Lortcher, and John Mecom and his wife, Mary Elizabeth. Mecom asked me when I was going home. When I told him that I was going the next day, he said: "I'll tell you what, I will come and pick you up in the morning. There is a fantastic toy store in town where you can buy fantastic toys for your children. You can buy a ten dollar toy for four dollars." I answered: "Uncle John (this was the way we addressed him), my children are grown up, and they do not play with toys anymore." He said: "What a pity. It is such a bargain."

Another unusual individual to whom Bill Thompson introduced me in Houston was Tuffy McCormick. Tuffy was probably six and a half feet tall, with a weight to match. He was huge. He was an All-American football player who had acquired the nickname of Tuffy by being really tough. Bill Thompson told me that he first met Tuffy when Bill was a young geologist working for the Standard Oil Company of California (now Chevron) in Midland, Texas. One day, Bill and Tuffy were on the same plane to Washington. Tuffy told Bill he was going to meet J. Edgar Hoover, the legendary head of the FBI. Bill thought that this was another of Tuffy's tall tales. When the plane landed in Washington, a large black limousine was parked on the tarmac. As Tuffy climbed down the airplane stairs, J. Edgar Hoover stepped out of the limousine. They shook hands, and both entered the limousine and drove away. Bill told me that, from that day on, he no longer doubted Tuffy's stories.

Tuffy tried to help me raise money. He was so well-liked in Texas and knew so many people that he almost succeeded. He could never remember the name I.N.O.C., and he called it "the Tel Aviv oil company."

Two experiences I had with Tuffy stand out in my mind. The first was when he arranged to have the Houston Museum opened after-hours, especially for "Dr." and Mrs. Alexander. The main benefactor of the museum, who was a member of the Rockefeller family, met us on the steps of the museum and personally showed us around.

The second experience was more amusing. We were all staying at the Warwick Hotel, that belonged to John Mecom. The Warwick, the best hotel in Houston, was named after Warwick Castle in England, from which the panels covering the lobby of the hotel had been taken. Bill Thompson and his wife, Byrdie, were in the presidential suite, Number

1110, whereas my wife, Rachel, and I had a nice room on the floor below. I had a meeting regarding Nigeria scheduled with Monsanto, near the Galleria, which is the nicest shopping area in Houston. After my meeting, we were all going to have lunch at the Warwick. When the taxi let me off outside the Monsanto office, I was surprised to see a very luxurious jewelry shop with the name Chazanov prominently displayed. This was very strange. Chazanov happens to be Rachel's maiden name, and there were only two Chazanovs in Israel, her brother and her cousin. Now I suddenly saw that name in Houston. I entered the shop and asked for Mr. Chazanov. Apparently they pronounced it differently and, noticing my accent, which was not exactly Texan, they became a little suspicious and refused to answer my questions. I went to my meeting with Monsanto and then returned to the hotel.

When we sat down to lunch, I related my experience at the jewelry shop. Tuffy McCormick called the waiter and asked for a telephone. He dialed and said: "Steve (this was Steve Chazanov). I have some folks of yours from Israel here. I want you to be at the hotel at five o'clock." Tuffy then suggested that we and the Thompsons switch our rooms for that meeting. We should take suite No. 1110 for that afternoon. This was a five-room suite, full of gold and gold-plated antiques, probably worth millions of dollars, which Mecom bought in Europe at the end of World War II. Sure enough, at five o'clock there was a knock on the door and Steve Chazanov walked in. He carried a small bag with the name of his shop printed on it. We spent a very pleasant couple of hours. There is no doubt that he was a distant relative of my wife's, as his facial features greatly resembled those of my wife's brother and cousin. As neither my wife nor he knew anything about their parents' and grandparents' birthplace in Russia, we could not establish the family link. After some two hours, Steve Chazanov left, taking back with him the little package he had brought in. To this day we do not know what was inside, but I have a strong hunch that when he heard his folks from Israel were visiting, he took a small trinket from the shop as a present. However, when he saw the "Middle Eastern oilman" in this five-room palatial suite, he was most probably embarrassed to give us such a modest present.

* * *

In May 1972, in New Orleans, Rachel and I were present at the ceremony of signing the Signal agreement with the Mecom interests.

After the event, I told Dutch Lortcher that I must speak with him privately concerning a very urgent matter. He said: "Fine. We (meaning myself, Rachel, Dutch and his girlfriend) shall fly on the Signal plane to Los Angeles. We can then talk privately on the plane." The rest of the Signal people who were in New Orleans would go back to Los Angeles by commercial flights.

During our flight, I told Dutch that as we had not yet raised the money, I was afraid that Signal might pull out of our deal. He said: "Don't worry, it won't happen." But I said, "I am still very much concerned."

After our arrival in Los Angeles, Dutch picked us up from the hotel for dinner. From the restaurant, he called Dr. Dinstein in Tel Aviv, Dutch told him about my concern and apprehension. He continued by saying that if the Israeli government would give a guarantee that it would furnish the company with a line of credit of three million dollars, if and when needed, Signal would close the deal. This money, provided by Israel, would be returned to the Government of Israel from the funds that were to be raised from new investors. Dinstein agreed.

One should remember that, although this was a very important decision on the part of Dr. Dinstein, it was not so unusual. To this day the Government of Israel gives very large grants to foreign investors, to encourage them to invest in Israel. In our case, PEDCO was still a 90 percent Israeli-owned company and the Signal investment was of very great importance to the country. Therefore, providing a line of credit to PEDCO was not out of the ordinary for the government.

Following the promised line of credit, the agreement with Signal was signed on the Island of Bermuda, and PEDCO began operations in June 1972.

Chapter Seventeen

Israel's Ministerial Committee Scratches its Head

The agreement with Signal and the creation of the Petroleum Exploration and Development Company (PEDCO), the Bermuda company into which I.N.O.C.'s concessions and Signal investment had to be transferred, needed the official approval of the Government of Israel. This approval had to be given by the Ministerial Economic Committee.

Instead of this approval being treated routinely, accompanied by a letter of praise and commendation from this committee, it became a tedious and dragged out process. The committee should have given me a "medal for outstanding achievement," praising the fact that an Israeli company had succeeded, for the first time in history, in obtaining a multi-million dollar investment from an international oil company. Finance Minister Sapir decided instead to make the procedure subject to a long process of presentation and lengthy approval. This, again, was an act of revenge for my daring to reopen the Israel National Oil Company which he had decided to close.

I had therefore to appear before the ministerial committee several times with full presentations to justify the merits of this investment. I had to explain why it was beneficial for the State of Israel to have millions of dollars invested by a foreign oil company, with the Israeli government retaining a 90 percent interest in the new company.

The situation was ludicrous, to say the least. The ministers were perplexed. If Sapir, who was so eager for foreign investments, subjected this investment to such a thorough scrutiny, there had to be something wrong with it. I had to meet, separately, eight ministers, to explain why it was "good for the Jews" to have Signal as a partner with I.N.O.C.

Several of these meetings will remain in my memory for ever. One was with Minister of Development and Tourism Moshe Kol. He was an elderly gentleman, very long-winded, and known for his not very impressive appearances in the Knesset, the Israeli parliament. He wanted to know: "Where are the five million dollars that you had promised to raise?" I told him that unfortunately my father was not a banker, and neither was his father, as far as I knew. I had been assured by Hayden Stone, who were old established bankers, that they were willing and able to raise the money. There was no reason to doubt their ability to do so. Therefore, I believed that they would deliver the goods.

The other memorable meeting was with Yigal Allon, the deputy prime minister. Yigal Allon, who was the commander of the "Palmach," the first and foremost fighting force of Israel before and during the War of Independence, was a legendary figure and one of the heroes of Israel. I went to see him, with Hanan Yavor, my Board member and friend, to Kibbutz Ginossar, on the shores of Lake Galilee, where Allon lived. He was a founding member of that Kibbutz. He received us very graciously and expressed his admiration for all our activities. Ethiopia was of particular interest to him, and he asked me to try and arrange an invitation for him to visit the emperor in Addis Ababa.

I met also with Shimon Peres who served then as minister of transport and who later became minister of defense and eventually prime minister. Peres received me coolly, almost frostily, and I did not understand why. He obviously understood very well the great political and economic importance of Signal's investment in I.N.O.C. Later I thought that Peres' cool reception could be attributed to my behavior upon my return from New York in 1957.

As I described in chapter 1, Peres, who in 1953 was the director-general of the Ministry of Defense, was the man who approved my posting to the army purchasing mission in New York in spite of the tremendous opposition to this appointment. Upon my return in 1957 I made a courtesy visit to his office. During our conversation he said: "Arrange your (private) affairs and come to see me" or words to that effect. This statement did not register with me and I did not understand that he probably wanted to offer me a position in the defense establishment. I did not return to see him for many years thereafter and he probably was offended by my ingratitude and did not forgive me.

Looking back it is quite possible that I had no comprehension of Peres' way of thinking and mode of operation. He was one of the leaders of a faction of Mapai, the ruling party, who were called "Ben Gurion Young Guard." The group included General Moshe Dayan, Itzhak

Navon, the future president of Israel, and other prominent young leaders. Against these "Young Turks" stood the old and established leaders of Mapai headed by the famous "Troika" (Triumvirat) of the minister of foreign affairs and the future prime minister of Israel—Golda Meir, Ziama Aran, the education minister, and Pinhas Sapir, the minister of trade and industry and the future minister of finance.

Peres as a political leader prepared a cadre of young and able men to assist him when the battle for control of the party and the country would take place after Ben Gurion's demise. As far as Peres was concerned I owed him a debt due to the fact that he had approved my posting to the very sought-after position in New York. Now, upon my return from the States, he probably planned to offer me a position in the defense department, alongside him. I not only ignored his offer but I joined the opposition by accepting my assignment with Lapidoth, under Mordechai Chen, who was Sapir's disciple—Sapir being the bitterest opponent of Peres.

This quite possibly might have been the reason for Peres' frosty reception. But, all this I understood only years later. At the time, all these considerations were completely beyond my comprehension, and I was not aware of the immense political struggles that were being fought behind the scenes.

The only minister who understood the great importance of Signal's involvement with I.N.O.C. was Jacob Shimshon Schapira, the minister of justice. During the ministerial committee meeting he stated that receiving a multi-million dollar investment in an Israeli oil company was an achievement to be very proud of. But I do not think that the other ministers understood it. They were confused and perplexed by Sapir's behavior.

Finally, two months later, after all of my meetings with the ministers, Signal's investment was approved by the Committee.

* * *

Several months after Signal's investment, Dutch Lortcher organized a meeting with Sumitomo, one of the largest Japanese conglomerates, to discuss the possibility of Sumitomo's investment of fourteen million dollars in PEDCO for a 40 percent interest in our company.

A delegation composed of four Signal executives—Dutch Lortcher, Peter Rainier V.P. Exploration, Bill Miller V.P. Legal Affairs, Richard

Bannister, Signal's geologist dealing with our projects and myself—arrived in Tokyo.

During three days of discussions it became apparent that the only stumbling block was Sumitomo's fear of the Arab Boycott. Signal offered to create a special subsidiary, which would hide Sumitomo's investment. They promised a reply within a week but unfortunately their apprehension of Arab response won the day. All that remained from this memorable trip was the photo of the lavish Geisha party that Sumitomo organized for us in Tokyo.

Chapter Eighteen

Drilling Offshore Ghana

1972 was quite a busy period. We were concluding the agreement with Signal, trying to raise money by private placement and then switching over to preparations for a public issue. We were also very active in pursuing new exploration opportunities in Africa, in the Far East, and in South and Central America, as well as in operating our existing licenses.

Although the agreement with Signal was signed only in May 1972, and that is when we received their investment of one and a half million dollars, we already felt from the end of 1971, that we had a "big brother" in Los Angeles and Houston. Signal's headquarters had moved to Houston in 1972. We could, and did, consult with them on each and every project that we wanted to review or investigate. We always received full support, advice and help. It was a different world altogether having a big and strong "brother" ready to come to our assistance whenever needed.

We sent out geologists to investigate possibilities in South and Central America. Efraim Aharoni, our chief geologist spent more than a month in Colombia, British Guyana, and the Dominican Republic. We checked possibilities in Ecuador and had other professional people travelling to the Philippines. In addition, we investigated new possibilities in Angola and other places in Africa.

On the one hand I wanted to enlarge our acreage portfolio, but on the other hand I was holding back subconsciously, having a feeling that our financial future was not secure as yet, and that we should not overextend ourselves before the foundations of the company were safe. Therefore, in spite of all these investigations, we did not apply for any new areas during that interim period.

During 1971 our Ghana group, composed of ourselves, through our wholly-owned subsidiary Volta Petroleum, Kewanee Oil, Ray Christian's Mayflower Company and Hamilton Brothers, drilled two deep offshore wells in Ghana, one in the eastern part, in the Keta Basin, which we named Keta I, and another one in the western part, which we named Tano I.

We had an excellent cooperative arrangement among the partners. Steve Cohen, our resident geologist in Ghana was managing the onshore operations, the geology and relations with the Government. We borrowed two professional men from Lapidoth who were put in charge of bookkeeping and supplies. The well-site geologist on the drilling platform was provided by Kewanee, drilling supervisors by Mayflower, and the overall drilling supervision by Hamilton Brothers. Their chief drilling superintendent, Don Holt, who was also supervising their drilling operations in the North Sea, was seconded to supervise the Ghanaian operations. This multi-national group worked in perfect harmony, and the two holes were drilled in record time without any mishaps.

The Keta well in the east of Ghana proved to be dry, but the one at Tano had some oil and gas which at the time was not considered commercial. The name we gave to the western area and to the well, Tano I, is still being used today, more than thirty years later. Subsequent exploration in this area by other companies discovered oil and gas in the Tano area. There is now a Tano Field and Tano South Field; thus our appellation was "immortalized."

In April 1972 two more participants joined our Ghana licenses, namely Mesa Petroleum and Diamond Shamrock Oil and Gas Company. Diamond Shamrock and Mesa conducted a detailed seismic survey in Keta, in the eastern part of Ghana, at a cost to them of approximately four hundred and fifty thousand dollars. They had a seismic option that was exercised by them in February 1973, to fulfill their commitment to drill a deep well onshore Keta at their full expense before June 30, 1973. They started drilling on May 21, 1973. Unfortunately, the well was dry.

* * *

I personally have not given up on Ghana. In 1979/80 I returned to Ghana, together with Russ Walker, a lawyer and oilman from Oklahoma City, who had a small oil company called R.J. Walker Oil Company. We decided to engage in exploration in Ghana. We sent Steve Cohen, the geologist who had worked for us in Ghana in the early 1970s, to try and obtain licenses in the name of the R.J. Walker Oil Company. Cohen spent a year in Ghana, but the Ghanaian government was under communist influence and no licenses were awarded. After a year of continuous efforts and an expenditure of a substantial amount of money, we terminated this effort. Russ Walker, by the way, was the son of Barth Walker, Mayflower's chairman and attorney, who had worked with us in Ghana in the early 1970s. We also had an illustrious partner in this venture. George W. Bush Jr, who was Russ Walker's roommate at Yale University, joined us for a small interest. Today, I can rightly claim that the President of the United States was once my partner!

I tried to explore in Ghana again in the mid-1980s, together with Ray Christian, the head of Mayflower and our original partner in Israel and Ghana. We sent Steve Cohen again to stay in Accra in order to try and obtain a concession. History repeated itself—the government's leftist leaning had not changed and, after a loss of several hundred thousand dollars, we gave up.

* * *

Signal joined us in Ethiopia and conducted a detailed seismic survey there at its own expense. Following this survey, we and Signal invited a serious company from Dallas, Texas, the General American Oil Company of Texas, to join us in Ethiopia. Signal and General American took the majority interest in the concession and undertook to drill two offshore wells at a cost of one and a half million dollars each, at no cost to us. We remained a 28 percent owner—at no cost.

Two extremely efficient wells were drilled in offshore Ethiopia by General American and their partners in the summer of 1973. Unfortunately, no commercial gas was found. The main attraction of offshore Ethiopia was a well drilled by Mobil Oil, in 1967, in which large quantities of gas were discovered. The well blew out. Mobil succeeded in controlling it, but then the Six-Day-War between the Arab countries and Israel broke out. Mobil abandoned the area.

A lot of internal discussions were held before drilling our two offshore wells in Ethiopia. One idea was to go to the original Mobil location and drill a twin well. After much consideration the partners decided to drill two new structures that were indicated by the seismic survey. Both wells were dry.

Our original concession area in offshore Ethiopia is still in demand thirty years later. New companies enter the picture all the time to join the search. As far as I know, none of them have had any success as yet.

* * *

Gabon was a different story altogether. This was a country with a large oil production. There were frequent announcements of new discoveries offshore Gabon. Had we been smart enough, we would have done a deal with Signal who were ready to drill two or three holes, for a 50 percent interest. They had a fair chance of being successful in one of those holes, as numerous oil fields were discovered on our blocks during the following years. The problem was that our head of exploration was greedy and demanded a hole on each of our five blocks. Unwisely, I did not intervene. Signal declined, and we lost the license five years later.

* * *

In Madagascar, Oxoco and its partners, Husky Oil Company from Canada, Kewanee Oil from Pennsylvania, and Invent Company conducted the offshore seismic survey that covered two and a half thousand miles and was completed in September 1971. The survey did not indicate promising areas to drill. Following the seismic survey, activities in Madagascar were suspended.

Chapter Nineteen

Oil Discovery in the North Sea

In July 1972, the Signal Consortium was allocated one block, in the northern part of the North Sea: block 211/18. Most of the seismic work in that part of the North Sea had been performed previously by various seismic companies at their own expense. They sold portions of the survey to the various oil companies applying for blocks in those areas. The applications for particular blocks were based on the interpretation of the seismic data that the companies had purchased previously. The oil companies that were allotted the relevant blocks could start drilling almost immediately, as the seismic picture was already clear to them.

Within one month, in September 1972, the first well was drilled by Santa Fe for the Signal Consortium. It was a discovery well, which flowed ten thousand barrels of oil a day. Unfortunately, all this oil flowed from one layer of sand which was quite thin. There were no additional layers of sand in the well. A single, and relatively thin sand does not make the well, or the area, a commercial discovery.

I remember sitting with Bill Thompson at the bar of the Warwick Hotel in Houston. He was then the executive vice-president of Signal. He became president less than a year later when Dutch Lortcher retired. I asked Bill what was going to happen in their North Sea block. Thompson took a napkin from the bar and drew on it a diagonal line, which represented a fault, saying: "If Laurence Snedden (Signal's chief geologist) is right, we have drilled on the 'upthrown' side of this fault which was to the west of the fault line. That side, being uplifted, was exposed to the elements millions of years ago, and any additional sands which might have been present in the section were eroded. If we drill on the 'downthrown' side, on the eastern side of the fault, we might find more sands. These sands may not have been eroded, and may contain additional reserves of oil."

Block 211/18

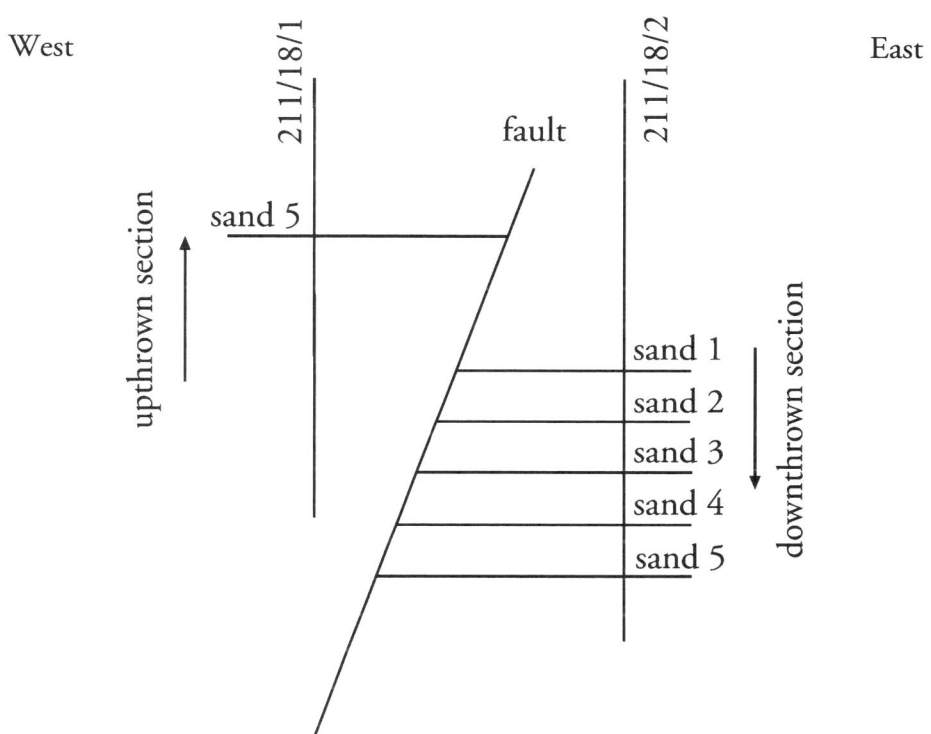

Drawing of the 211/18 fault

The second well, that commenced drilling in June 1973, revealed four more sands on the eastern side of the fault. A major oil field, the fourth or fifth largest in the North Sea, was discovered. This was the Thistle oil field, which eventually produced eight hundred million barrels of oil.

As mentioned before, we had a secret participation of a 5 percent economic interest in this discovery. In the next chapter, the saga relating to this interest is described in detail.

There are two basic truisms in the oil industry. One states that oil is found in the minds of men; the other is that oil is where you find it. The conversation with Bill Thompson in the bar of the Warwick Hotel and the subsequent discovery of the Thistle oil field proved both statements to be true.

* * *

The Norwegian side of the North Sea proved to be more productive than the British one. Signal was eager to enter the oil exploration offshore Norway as well. In order to be considered favorably by the Norwegian government, Signal was instrumental in creating a local Norwegian oil company with which Signal would be the technical leader. On one of their trips to Norway, Dutch Lortcher invited my wife and me to join him. We met in the port of Bergen, which was the headquarters of the Norwegian oil industry. In three days, we must have attended four parties, each hosted by one or the other of the Norwegian investors. The Scandinavians drink like fish, but the fellow who picked us up at the hotel did not put a glass to his mouth. I asked him whether he did not like drinking. He replied that, as he was driving, he did not dare to touch liquor. The Norwegian police constantly checked drivers on the road with a breath analyzer. Any driver whose blood showed an alcohol content larger than the prescribed limit was removed from his car and taken straight to jail. There was no waiting for judges or court cases, etc. The car was impounded and the driver had to stay in jail for a week or so, to sober up. This procedure almost eliminated drunken driving in Norway.

On our way back, we stopped in Oslo, where we stayed for another two days. We were taken for a drive around town by one of the Norwegian participants in the oil group. When we came to the edge of the city, we saw an abandoned farm. Our Norwegian host explained that when the town grows and reaches farm land, the farm is confiscated and the farmer is given new land further away. The rationale behind this system is that the farmer was not responsible for the rise in the value of his land; therefore he should not profit from its conversion to a building site. Although I am very far from being a socialist, I think that this is a very wise and just law. When one sees all the farmers in Israel becoming mega-rich due to the fact that their agricultural land is turned into building sites, the wisdom of the Norwegian law becomes quite apparent.

If it was in existence in Israel, such a law would have even more justification. Almost all agricultural land in Israel was not purchased by the settlers but was given to them either by the Jewish Agency or the government itself.

* * *

A few months later I wanted to investigate the possibility of applying directly for an an oil license in Norway. The Israeli ambassador encouraged me and arranged a meeting with a senior official of the relevant Norwegian ministry. The meeting was scheduled for twelve noon in the ambassador's residence. The ambassador warned me that Norwegians are very punctual and that I should be there before the appointed time. We arrived ten minutes early and entered the ambassador's residence. Outside, in the snow-covered garden, we saw a big man bundled in a heavy coat and warm hat, walking back and forth. We were convinced that this was the gardener.

At twelve noon, on the dot, the "gardener" walked in. He was, in fact, the man whom we had come to meet, the high official of the Norwegian Oil Ministry. The Scandinavian "disease" of punctuality did not allow him to enter a minute earlier!

After a pleasant discussion, it became clear to us that we would not pass the scrutiny of the Norwegian ministry. We were too small and too poor, to name only two of the main obstacles. It was a nice try.

Chapter Twenty

Public Share Issue and Arab Boycott Again

After the agreement with Signal was signed, we started to plan a public issue of the company's shares. We went again to Hayden Stone. They invited a San Francisco-based investment house, named Bateman-Eichler-Hill-Richards, to join them.

It was only when we started preparing the prospectus that I realized how costly such an exercise was. I was sitting together with four lawyers, two from Israel and two from the US, all representing our side, plus additional lawyers of the underwriting bankers, plus several accountants. Each hour thus cost more than a thousand dollars. To illustrate the futility and wastefulness of such an undertaking, one example will suffice. They were all arguing, for almost three hours, as to how to describe the fact that the waters offshore Ghana were shallow and that the sea was calm. Both of these facts indicated that the drilling offshore Ghana would be relatively easy and not very expensive. They were afraid to be too positive in their description, because a sudden, unexpected storm might blow up in those calm waters. The risk was that the company, the bankers and their lawyers might be sued by a litigious investor claiming that the prospectus was misleading.

Due to the immense power of the Securities and Exchange Commission which regulates all public companies, on the one hand, and the "eagerness and enthusiasm" of American lawyers to sue anybody and everybody in sight, on the other, a prospectus became an insurance policy and not a selling document. If an investor read all the risks and warnings appearing in any company's prospectus, he would surely never invest in it. Luckily very few investors read all those warnings. The decision of the investor is usually based on the reputation of the company, its product, its management, and the forecast for future profits. Still, the most important factor is the "mood" of the market, which, by the way, is almost never neutral. It is either euphoric or depressive.

* * *

During one of the discussions, the underwriters' lawyers told me that they would have to go and visit each country in Africa where we had concessions. They had to check the validity of our rights, otherwise they would not be able to sign their name on the prospectus.

I was furious. In addition to the waste of time and money such trips would entail, I was afraid that the relevant African ministries, once again seeing new white faces, would become very suspicious. Some months earlier, they had received the Signal representatives checking the same set of papers. There must be something very wrong with this company if it is being investigated every six months.

The lawyers were adamant. They must go, otherwise they would not be fulfilling their legal duties. Suddenly I had a brain wave: "Go ahead," I said. "You may go any time you want. But I have to warn you, the moment you land in any of those countries, you will be arrested and put in jail. Not for long, probably not for more than four or five days, since actually you did not commit any crime. They will explain to you later that this was a small misunderstanding; but I assure you that four or five days in an African jail will be a lifetime experience for you."

"You do not mean that," they said.

"I certainly do, I was never more serious in my life," I replied. "Surely you can have no doubt that we have enough good friends in those countries who will oblige us with this small favor." As if by magic, their desire to travel to Africa and the necessity and urgency of their trip vanished. The subject disappeared from the agenda.

We had another fight on our hands, this one with the US accountants. Their "conscience" did not allow them not to mention some of the "questionable" payments we made in Africa. The fact that this was the only way to do business in many countries in the developing world did not interest them. I do not remember now how we overcame this obstacle, but again it was an unnecessary waste of money and nerves.

A prospectus is usually printed at night. Each and every change in the day-to-day life of a company affects the story in the prospectus, and a new version of the prospectus has to be drafted and printed. In the end, our printing bill came close to a hundred thousand dollars.

* * *

I had one pleasant experience during that trying period which taught me a lot about how money could and should be raised. I was told that one of the bankers of Bateman Eichler, the co-underwriters of our public issue, was visiting London. They suggested that it would be a good idea for me to meet him. His name was William N. L. Hutchinson Jr, or, in short, Bill.

Bill was a member of a very old and prestigious family in San Francisco. Both his grandfather and his father had been generals in the US Army, and he himself had been a colonel in World War II. When he started his banking career, he decided to find a niche for himself and chose to concentrate on England. Commercial ties of the American banking community with England were almost non-existent after the war. Bill started travelling to England frequently. He succeeded in establishing excellent relations, and a feeling of trust and great respect, with many investment managers in the City of London.

I went to London to meet Bill Hutchinson. He told me that he had received from his firm a draft of our prospectus but had not yet read it. He complained that, if his firm wanted him to do a decent job, they should have called him and explained the whole matter to him. Ours was not a regular issue; Israel, oil exploration, Africa—it all needed explaining. This was not an issue of shares of a company manufacturing steel or making television sets.

I spent a whole day with Bill, outlining what we were doing and how we were going about it. Bill liked our story very much and became enthusiastic. In the next few days, he took me to visit ten financial institutions in the City of London. Six of them each committed to invest two hundred and fifty thousand dollars, their only condition being that the prospectus become effective and the company be floated on the public market. The roster of institutions that agreed to invest was very impressive. One was Swiss Bank Corporation, the second largest bank in Switzerland, and another was Legal and General, one of the largest insurance companies in London. Hill Samuel, a very prestigious merchant bank, was the third participant. Even the institutions that declined did it with grace. I observed with admiration how Bill was received, how well he had studied the material, how he presented the company. It was a pleasure to listen and to learn from him.

Bill's wife was a member of a very prominent Italian family, and he was one of the Vatican's financial advisors. The Vatican gave very serious consideration to making an investment in PEDCO. In the end they had to decline due to political considerations.

During that one week in England, Bill Hutchinson succeeded in getting commitments for one and a half million dollars. This was 25 percent of the entire public issue. I was convinced that, if Hayden Stone or Bateman Eichler had other "Bill Hutchinsons," we were going to raise the money without too much difficulty.

Following my experience with Bill Hutchinson in London, I told the underwriters in one of our meetings: "You spent hundreds of thousands of dollars on our issue. Your lawyers are not cheaper than ours. Each of you keeps two people, full time, dedicated to this project. If you add it all up, you arrive at a very substantial sum of money. For the life of me, I cannot understand why you do not make a list of people or institutions that such an investment will appeal to. It should then be presented, on a one to one basis, to those prospective investors. Instead of making such a list and approaching investors individually, all you do is send out the prospectus. This prospectus is one of a hundred prospectuses that investors receive every month, and most of them end their life in the wastepaper basket."

I told them a story that I had heard from my professor of marketing at Columbia University Graduate School of Business. He was one of the sharpest individuals I have ever met in my life. The students hardly dared to ask him a question, for fear of receiving a scalding reply. He taught us that there were three cardinal rules of marketing. The first one was that you cannot sell the same product to everybody. Some products suit rich people, while others appeal to the poor. Therefore, the second rule of marketing is to find the segment of the population that you should aim at. The third rule is to identify the segment of the population in which you are most likely to succeed. He then told us an interesting story from his own past experience. After World War II, he and some other professors, including one from the school of dentistry, decided to go into business.

The dentistry professor invented a revolutionary dental tool. The professors decided to market the instrument and make a fortune. Each of them invested all his savings in this project. They printed a beautiful brochure describing the wonderful invention, and sent it out to thousands of dentists. Nothing happened. Not one dentist responded.

"Where did we go wrong?" asked our professor. The answer was that they had not remembered the most important rule of marketing, that one must find the best way to reach the prospective customer, to get the dentist's attention. He continued: "We forgot that each dentist has a secretary. Many brochures, each describing different products and inventions, are received daily. The secretary throws them all into the waste basket. The dentist himself does not have the time or inclination to read all this advertising material. The most effective way to reach a dentist is through a personal visit and through articles in a professional dental journal. The product should also be presented at a dentists' convention. That was the right way, and the only way, to market the invention."

The professors lost their savings and decided to concentrate on teaching.

I tried to convince the underwriters to draw the conclusions from the lesson I learned from my professor. I added: "Look, among your many clients there must be some to whom our investment might appeal. You have to sit down and figure out who is not afraid of the Arabs, who is interested in oil and other natural resources; who has a gambling streak and is not averse to risk taking; who has uncommitted funds; who has made a lot of money recently in a new and unusual project. You have to approach all these people on an individual basis." It did not work. I did not succeed in convincing them—either they did not have the right people to make such an approach, or the slump in the market had affected them so badly.

At the beginning of May 1973, the stock market was in terrible shape. At one stage, probably in March or April, the underwriters wanted to postpone the issue. I flew to New York for forty-eight hours. Signal's people arrived from Houston for that meeting. We met with Sandy Weil, the head of Hayden Stone who is now the head of Citigroup, the largest and richest financial organization in the world, and forced him to continue with the public issue. I told him that we must have the money and we must have it now. If they tried hard enough, they could sell the issue. "Look at what Hutchinson accomplished in Europe in one week: why can't you do the same?" He agreed to persevere.

While all of this was going on, another big problem cropped up. Equity Funding Corporation, the insurance company which was our partner in Ethiopia and which had supported us financially for almost five years, was going under. For five years they had paid our overheads, our geological fees and other related expenses. Suddenly, in March 1973, they were discovered to be crooks. They had stolen hundreds of millions of dollars from the public. The president of the company and other senior officials received long-term jail sentences.

The Equity Funding saga in itself is an interesting story. They were an insurance company engaged in selling to the public a life insurance policy together with participation in a mutual fund. The idea was that the profits and dividends of the mutual funds would suffice to cover the cost of the insurance policy premium. Thus the investor received two benefits for the price of one. For years in the late 1960s, this system worked well and the company grew by 30 percent each year. The company's stock was soaring. One year, probably in 1970, the profit forecast was poor. Management was afraid that the price of their shares would drop considerably. They decided, secretly, to create new profits from people whom they "killed," on paper, and to collect their life insurance from re-insurers. To choose the innocent "victims" they picked the names, at random, from a telephone book. This system continued peacefully and profitably for several years. Unfortunately, in 1973, they fired one of their senior staff who went to the authorities and told them the reason for the continued "success" of Equity Funding. The company declared bankruptcy and several of their executives went to jail.

In order to eliminate their name from the prospectus, we had to buy back their interest in the Ethiopian venture. Although we did not have to pay them any money, each such change necessitated rewriting the prospectus. This change was another indication that the sooner we could come out into the market with a finished prospectus the better.

While all this was going on, we were faced with a new battlefront. Santa Fe found out from the various published documents, which had been distributed among many investors, that I.N.O.C. was still involved in the North Sea consortium, albeit in a round-about way. As may be remembered, PEDCO's obligation was to invest 5 percent of the cost of exploration and development of Signal's North Sea block 211/18. PEDCO's right was to have the economic interest attributed to such 5 percent investment. With the resulting income, PEDCO was permitted to buy, from Signal, 5 percent of the oil production of that block.

Santa Fe was livid. They went to see the parent company of the Signal Group, complaining that Signal Oil was destroying twenty years of friendship between the two companies. Both were based at the West Coast and had had very good relations and many successful deals in the past. They claimed that Signal Oil might cause Santa Fe irreparable damage. Almost all Santa Fe's drilling equipment was in the Arab world.

Signal's action was putting these interests in jeopardy. They complained that Signal Oil had concealed from them the fact of a continuing Israeli involvement in the North Sea block, in spite of the reassurances, given by Signal, that the Israeli involvement no longer existed. Another legal complaint was that the original consortium agreement limited participation in the group to companies with a net worth of at least twenty million dollars. PEDCO did not meet this criterion, another breach of the agreement.

Santa Fe was very serious. They were threatening with a law suit claiming damages of one billion dollars. It was an immense sum at the time, and even now, over thirty years later, it still is a very large sum of money.

The "war" became ugly. Signal, in one of their letters, used the argument that anti-Semitism was contrary to the US law. There were four groups of lawyers involved in this dispute: Signal's lawyers, our lawyers in America—Ken Bialkin of Wilkie Farr, and our lawyers in Israel from Adi Ephrat's office. The lawyers representing the Government of Israel joined the fray, as well. There was a danger that Santa Fe would go to the Securities and Exchange Commission in Washington and demand an injunction, claiming that the economic rights in the North Sea, described in our prospectus, had been illegally obtained. These rights were now being challenged. Therefore the prospectus should either be scrapped or the whole North Sea chapter, and the value attributed to it, removed from the prospectus. Obviously such a step would have killed the public issue immediately.

The Santa Fe debacle came up for discussion before the Israeli Cabinet during several sessions.

One has to remember that this dispute took place in the summer of 1973, before the drilling of the second Signal well, on block 211/18. That well eventually confirmed the existence of an oil field. It was therefore not clear, at the time of the dispute with Santa Fe, whether Signal had a commercial discovery and whether the big fight was worthwhile.

Finally a compromise formula was suggested by Signal. PEDCO would relinquish its right to purchase 5 percent of the oil production. It was the only solution which had a chance to pacify Santa Fe and allow PEDCO to proceed with the prospectus. Signal strongly urged us to accept this compromise. They were afraid that we might lose everything, including the possibility of proceeding with the prospectus, and raising the money. Signal added privately that they would be happy to sell us oil.

But our legal right to purchase part of the North Sea production had to be eliminated from the official documents.

Signal's suggestion was brought to our Board of Directors and then before the Israeli Cabinet which decided to accept Signal's proposal.

It should be remembered also that our 5 percent participation in the North Sea was a present given to us by Signal. We could not have gotten it through any other means. Also, without Signal's participation, we could not issue the prospectus. It would have been worthless. If we did not agree to Signal's proposal, there was the danger that Signal would decide to write off its investment in PEDCO and forget this complicated and unpleasant chapter of their involvement with I.N.O.C. that continued to create unnecessary problems for them.

Another danger was the possibility of losing Signal's support and participation in Gabon and Ethiopia. The only logical solution, therefore, was to accept Signal's solution to the North Sea problem.

Chapter Twenty-one

The Road Show

The underwriters and their lawyers had one amusing dispute with the Israeli Government. This dispute concerned my salary, which had to be mentioned in the prospectus. My salary at the time was something like four hundred dollars a month, a sum which was considered to be a joke in the United States. The underwriters claimed that the caliber of the management in the public eye is judged by the salary of its executives. If the prospectus named my true salary, the prospectus would be laughed at. Obviously, no sane investors would touch it. This was the time that salaries in Israel were still based on the anachronistic Labor Union rule that fixed the salary of the janitor on the same scale as that of the president, and if the janitor had more children than the president, he was paid more. There were exceptions to this rule, but not many and not to a great extent. After a long discussion between the underwriters' lawyers and Dr. Zevi Dinstein, the "oil czar" of Israel, it was agreed to increase my salary to the princely sum of one thousand dollars a month. This salary was also going to be linked to that of the highest paid public official in Israel, who was then the president of the oil refinery in Haifa. I remember Dinstein stating that, if my salary would be fixed at a level higher than one thousand dollars a month, he would have no choice but to cancel the public issue. In order to overcome this problem, it was decided to lump all the salaries of myself, the company directors and the senior personnel into one sum, to prevent it from looking ridiculous in the prospectus.

In April 1972 our daughter Shaula got married. As both she and her husband were studying for their master's degrees in physics and mathematics at The Hebrew University in Jerusalem, we tried to find an inexpensive apartment for them. We found one for the price of

IL90,000 (Israel pounds), which at the then-prevailing rate of exchange was approximately $21,000.

The young couple had about IL15,000 saved from their wedding presents. I received a loan of IL30,000 from I.N.O.C. There was also a special mortgage available for newlyweds of something like IL30,000. This mortgage had to be approved by a Jerusalem public committee, which administered the granting of such mortgages. I phoned Kenneth Bialkin, our lawyer in New York. He was a good friend of Teddy Kollek's the famous mayor of Jerusalem. I asked Kenny to talk to Kollek and recommend that Shaula be granted this mortgage. I had been acquainted with Teddy Kollek for many years, but I also knew that he could not refuse Bialkin's request, whereas he could refuse mine. After Bialkin's call, I phoned Teddy Kollek. He was quite upset. He said: "You, the president of Israel National Oil Company, need to avail yourself of a mortgage that is designed for poor newlyweds?" I answered: "Teddy, I already borrowed as much as I could and they need this mortgage." Probably he could not understand that in my position I really had no additional sources of finance. He finally approved the mortgage. I often wondered why he was so convinced that I should have had the money.

Now, thirty years later, the salaries of professional people in Israel are comparable to the salaries of similar professionals in the States. In some cases they surpass them. But this is now and that was then.

* * *

In June 1973, the final prospectus was being prepared. We hoped that the dispute with Santa Fe had simmered down and we would be able to issue the prospectus and raise the money. As described earlier, the discovery of the Thistle oil field in the North Sea, was still four to six weeks away.

Such a prospectus is called a "Red Herring" because it has red notations all over its cover, indicating that this is the final version but one. What is missing is the date, the price per share, and the total sum which had to be raised.

This "Red Herring" document is printed in thousands of copies and is distributed to all the offices of the underwriters. What remains to be done is a "Road Show" where the issuing company and the underwriters present the company to the brokers of the underwriters and answer questions. Finally, when the underwriters are convinced that they have effectually pre-sold the issue, the prospectus becomes effective.

In a way, the word "underwriter" is misleading. Underwriters underwrite an issue, which means that they bought it. They only do it when they are convinced that they have pre-sold it. Their exposure is minimal, usually as short as half a day between the issue becoming effective and the printing of the final prospectus. The following day the issue is sold to the public. Even in such a case, when the interval is less than twenty-four hours, "disasters" to the underwriters do happen. While we were dealing with our prospectus, there was an issue of bonds for IBM with a fixed rate of interest. The issue became effective on a certain date and IBM was paid by the underwriters. The same evening, the Federal Reserve increased the rate of interest, and the underwriters suffered heavy losses by having to sell the bonds to the public at a lower price than they had paid IBM.

In July we were invited for the Road Show. I had to convince my wife Rachel to join me. She did not want to go, saying that whenever she went with me some disaster happens. First there was the Patino story mentioned in Chapter Seven, then the Nigerian disaster, followed by the death of the president of Charterhouse, and so on and so forth.

I finally persuaded her to join me by promising that we were just going to travel from town to town, for six days, at the end of which I would collect a check for six million dollars. I promised her there was really nothing to be concerned about. It was going to be a pleasure trip. Unfortunately, as in many other cases, she was right.

We first went to Los Angeles, where Bateman Eichler assembled more than a hundred of their salesmen at their offices. We sat on the dais, answered questions, showed maps and smiled. It was difficult for me to judge the mood of the salesmen and their reaction.

The next day, we went to San Francisco and the story repeated itself. Bill Hutchinson was present at that meeting. He said that he was afraid the issue was not going to succeed and reminded me of the way he had obtained the one and a half million dollar commitments in London. This was the only way, in his view, that our issue could be placed with investors in the United States. The underwriters should have identified a small number of venturesome private and/or institutional investors who might have an appetite for such an issue. They should have held individual meetings with each such investor explaining the complexity of the issue and the very high possible rewards of such an investment. In Bill's view, in spite of the doldrums in which the market found itself, the underwriters could have placed all of the issue or possibly even be oversubscribed. Instead, they continued with their mass marketing method, which would most probably fail in the then-terrible market conditions.

The next day we flew to Miami, Florida. When we walked out of the airport at 2 a.m., it was like walking into a blazing furnace. In the morning we went to the Hayden Stone offices in Miami. The head of the office was absent. I was told that he had a "doctor's appointment." When I saw that the man in charge was not present, I felt a disaster was imminent.

I had a long conversation with one of Hayden Stone's brokers in that office. He explained to me that all a securities salesman has is a card index of the names of his customers. This is his stock-in-trade. When times are good, he calls his customers every day. They call him, and everybody is happy. When times are bad, the salesman dreads each new telephone call. It is probably one of his customers complaining that the stock he bought the other day was going down. The salesman is not very much concerned about his employer, which in that case was Hayden Stone. What is precious to the broker is his list of customers. He could move the next day to another investment house, and his customers would probably follow him to the new place. Therefore, the last thing he would like to do, in a bad market, is to offer his investors a new issue of a speculative venture.

That evening, we received a call from Hayden Stone's headquarters in New York that they were withdrawing the issue.

Chapter Twenty-two

The 1973 Yom Kippur War

At the beginning of August of 1973, I had a long meeting in New York with Zeev Sher, the economic minister of Israel in the United States. Zeev was very familiar with our situation. He was formerly the deputy attorney general of Israel and, as such, participated in all the economic ministerial committees when they discussed the Signal investment. He also represented the Israeli Government in the signing of the two Signal agreements. When he was posted to the States, he closely followed our public issue efforts.

Zeev Sher was against our public issue from the very beginning. He was also seriously concerned about the new attack by Santa Fe. He rightly claimed that you cannot, on the one hand, conduct oil exploration efforts around the world under various guises and, on the other, "undress" completely, and publish everything openly in a prospectus, in order to comply with SEC rules. He strongly voiced his opinion that the Government should finance I.N.O.C.'s exploration efforts. In his view, our activities were very important to the welfare of the State of Israel, both politically and economically. But this was his personal view, and it had no influence on the policy of the Ministry of Finance and its all-powerful dictator, Finance Minister Pinhas Sapir.

During our meeting, which took place soon after the failure of the public issue, Zeev Sher voiced his concern about the Santa Fe problem. He also mentioned Israel's deteriorating position in Africa and the negative effect this was going to have on our concessions.

He told me that he was going to write officially to Minister Sapir and would recommend that the Government seriously consider the sale of all I.N.O.C.'s assets. In his letter, which he wrote soon thereafter, he voiced his concerns and stated that time was working against us. He urged the

Government to act quickly, before irreparable damage might be caused to
I.N.O.C.'s properties, both by Santa Fe's actions and the continuing loss
of Israel's standing in Africa.

During the same week, in New York, I met with Max Fisher with
whom I had had previous dealings. I have mentioned previously that
Fisher represented Signal's secret ownership of Paz during the 1960s, and
that in 1969 he offered to finance all our foreign operations. Max Fisher
was a very wealthy oilman, the treasurer of Richard Nixon's presidential
campaign, and one of the heads of the Jewish community in the United
States. He was very much involved in Israeli affairs. Fisher came
independently to the same conclusion as Zeev Sher. He met with Sher
and, following that meeting, Fisher wrote his own letter to the Israeli
Government recommending the sale of I.N.O.C.'s assets.

My feelings were ambivalent. On the one hand I understood the merit
of selling and the benefit of such sale to the Israeli Government, which
would receive many millions of dollars for our holdings. The risk of a
possible loss of some of those assets did not escape me. On the other
hand, to sell something that I had created, starting from nothing or rather
from a negative value, and that had become a very valuable property, was
heartbreaking.

There was also a personal consideration. If the company would be
sold, I would lose my position and the work I loved.

I thought that we might have one more alternative. I told Dr. Dinstein,
when I returned to Israel, that we should consider the possibility of living
through the present "hurricane" by lying low and trying to survive, hoping
that the sun would soon come out again. The Santa Fe debacle might pass,
the situation in Africa might improve, we might be able to redirect our
efforts towards Central America, and so on and so forth.

Dinstein presented the various alternatives to Minister Sapir and then
to the full Cabinet, and the decision reached there was to sell. I was
informed accordingly.

While these deliberations were going on, I developed the idea that, if
we could find an independent buyer, who had no relations with, or fear
of, the Arab world, and who wanted to enter the energy business, Israel
might still retain a safe source of supply of oil. Israel might reach an
agreement, secret or open, to purchase from such a buyer any oil
production which this buyer would develop from I.N.O.C.'s assets.

I pointed out to Dinstein that if we sold out to an oil company our
assets would be swallowed and merged with the purchaser's assets, with

no trace left of our holdings. In such a case, the simplest selling procedure would be to engage a merchant bank to find the best buyer, and to accept the highest bid. On the other hand, if an independent buyer could be found, we might still retain rights to purchase oil. Also, such an independent buyer would most probably want to develop the company further. He would thus create more assets, and the benefit to the State of Israel of having an assured supply of oil might be even greater.

The subject of an asking price was quite simple. The value of the company had already been established both by the public issue and the Signal investment. Signal paid one and a half million dollars for 10 percent. Therefore fifteen million dollars for 90 percent of the company would be a logical and reasonable figure. Our demand for a right to purchase oil from the new owner of the company might reduce this price.

Dinstein was very skeptical about both of my assumptions, i.e., the possibility of finding an independent buyer and of achieving the fifteen million dollar target price. Still, he allowed me to go ahead and try.

During September I already had serious meetings in London with the House of Warburg. They were planning to create an oil "conglomerate" by putting together an oil exploration company, a service company, and an oil supply company under one roof. We were being considered to be the oil company in that conglomerate. We also had a series of meetings with a small banking house named Brandt and other meetings with a Canadian investment company.

I remember attending a party in London on Thursday, October 4, at the home of Ananda Krishnan, a Malaysian oilman, who was also helping me to identify a possible buyer for our assets. Oilmen from various countries were present at the party, including an Egyptian oilman, with whom I had a pleasant and lengthy conversation. The next morning, Friday, October 5, I left for home.

I arrived home late Friday afternoon. My son Kobi, who was then serving in the Army as a deputy intelligence officer of the Jerusalem Brigade was there. As this was the eve of Yom Kippur, the holiest day of the Jewish calendar, I asked my son whether he would join me in the synagogue. He replied that he had to be near a telephone, as his superior officer feared that war was imminent. When I returned home from the synagogue, my son was already on the way to Jerusalem to join his unit. He had been called up while we were in the synagogue. The next day, at 2 p.m., Israel was attacked without any warning by Egypt and Syria on two fronts, in the south and the north. The Yom Kippur War had begun.

It was a terrible war. Israel suffered thousands of casualties. For several days it was touch and go, and the very survival of the State of Israel was at stake. Several of my friends lost their sons fighting on various fronts. It was a time of uncertainty, grief and disillusionment. In the second week of the war, Israel went on the offensive and at war's end found itself on the western side of the Suez Canal, one hundred kilometers from Cairo, as well as twenty-five kilometers from Damascus, on the Syrian front. But the price was much too high. The lack of preparation of the Israeli forces and the complacency of the Israeli Government were unforgivable.

Following the Yom Kippur War the rest of the African countries severed their diplomatic relations with Israel. Our company was completely immobilized with regard to our African holdings. The Arabs became the kings of the world. They declared an oil embargo, claiming that the United States had helped Israel during the war. The price of oil quintupled. It was a new world altogether.

* * *

In November, we renewed our efforts to find a buyer for the company. One of the first people with whom I discussed such a possibility was Meshulam Riklis. He was an Israeli who had emigrated to the States many years earlier and had made a large fortune there. He was a controlling shareholder in a number of large public companies. Riklis had graduated from the Herzliya Gymnasium one year after I did, and we knew each other slightly.

I met Riklis again at the beginning of 1973 in New York at a meeting of ADL—the Anti Defamation League—probably the most influential and strongest Jewish body in the USA. On the agenda of that meeting was the growing Arab power, due to their control of oil supplies. Ken Bialkin, our lawyer in New York, invited me to participate. The only speaker who recognized the Arab danger was Meshulam Riklis. I seconded his statements and told the assembled group of Jewish leaders that the danger was very serious and growing every day. After the meeting, Riklis and I spoke briefly together on the subject.

In view of the above, I felt that he might possibly be willing and able to buy I.N.O.C.'s interest in PEDCO. I called him and he asked me to come to Florida, where he was vacationing. We spent half a day discussing the matter, and then I flew back with him, on his company's plane, to New York.

The following day he instructed one of his executives to meet with me and pursue the negotiations. The man phoned Hayden Stone, in my presence, and received confirmation of the various facts I had given him. I remember his reply to a telephone call he received during our meeting, telling the caller: "I am busy, we are buying an oil company."

After several more meetings, Riklis told me that he was coming to Israel and would finalize the purchase price with Sapir. Riklis arrived in Israel accompanied by an airline hostess and was incommunicado for the entire length of his visit. He spent four days locked up with his companion in his hotel suite. I begged Dinstein to have Sapir call him, but to no avail. This was the end of our "love affair" with Riklis.

During one of our Board meetings, probably in December, I was attacked by three members of the Board, including the head of the Government Companies Authority, Jacob Salman. They claimed that my demands were exaggerated. By asking for fifteen million dollars, which, in their view, was an exorbitant price, I was destroying all possibilities of selling the company. They urged me to reduce the price to ten million dollars. They feared that we would end up without anything. I remember saying that I did not feel that the price was the determining factor. If we should find a suitable buyer, who would agree to have continued relations with Israel and to supply us with oil, whether directly or indirectly, the price would not be an obstacle. We had not met anybody yet who expressed a willingness to consider the purchase, but at a lower price. The Board gave me an additional month of grace to find a buyer.

The situation at Signal also changed dramatically. There was a take-over attempt on Signal Companies, with the prize being Signal Oil. In order to forestall this attack, Signal Companies sold Signal Oil to Burmah Oil at the end of 1973. Signal Oil became Burmah Oil USA on January 1, 1974. Suddenly, instead of the friendly shareholder in PEDCO, we had the formidable Burmah Oil as a shareholder whose interests lay completely in the Arab world. Selling the company became of much greater urgency.

* * *

In late November 1973, in London, I received an urgent call from Bill Thompson, the president of Signal. Bill asked me whether I was standing or sitting. I replied that I was standing. He told me to sit down and listen.

Bill said that Signal was being attacked by Mark Millard, the Loeb Rhoades banker. As may be remembered, Mark Millard was the first investment banker I met in New York, back in 1969, during my initial attempt to raise money on Wall Street. I have described my meetings with him in Chapter Eight.

Millard represented the Bronfman family from Canada, the owners of Seagram, the whisky company. The Bronfmans also owned an oil company, Union of Texas Petroleum. They wanted to take over Signal Oil and Gas and combine it with Union of Texas Petroleum. The Board of Signal Companies, the parent of Signal Oil and Gas, authorized Bill Thompson to approach the Israeli Government with a proposal that the Government of Israel directly or indirectly should buy Signal Oil and Gas Company. Another director of Signal Oil was on his way to Kuwait to propose the same transaction to the Kuwaiti Government.

This was a once in a lifetime opportunity for Israel to solve its oil supply problems. I called Joe Boxenbaum, our Board member, who was a friend of Sapir's and who previously represented Signal's shareholdings as a member of the Board of Directors of Paz Oil in Israel. I related to him my conversation with Thompson. Joe went to Sapir with the proposal. In turn, Sapir consulted with Golda Meir, the Prime Minister of Israel. The idea was too big for them. They could not think in global terms. One has to say, in their defense, that this offer came at a very bad time. It was not long after the terrible Yom Kippur War had ended. Golda Meir was attacked for being too complacent before the joint Egyptian and Syrian attack suddenly started. She and her colleagues in the Cabinet were accused of being among those responsible for Israel's not being properly prepared. She resigned from the Government four months later with a broken heart. This was not the time to suggest to her to buy an international oil company.

Chapter Twenty-three

The Sale of the Company to a British Bank

After Burmah Oil bought Signal, they became 10 percent shareholders of PEDCO. Bill Thompson became the president of Burmah Oil's American subsidiary, and thus, at least, one good friend of ours remained in the management of Burmah Oil. They started pressing for Israeli involvement in PEDCO to cease as soon as possible. They were concerned that their partnership with an Israeli Government-owned company might harm them in the Arab world.

Sometime in January 1974, we were finally introduced to an interested buyer. The introduction was made by Ananda Krishnan, the young Malaysian businessman whom I first mentioned in chapter 21. He lived in London and was involved in oil exploration in Indonesia. Ananda had an M.B.A. degree from Harvard Business School. He was a very energetic and able young man. Today, in 2003, he is a multi-billionaire, and one of the wealthiest entrepreneurs in Kuala Lumpur, the capital of Malaysia. He is the man behind the erection of Petronas Towers, the tallest buildings in the world.

Ananda Krishnan was very friendly with Allen Challis, the deputy head of the Crown Agents in London. The Crown Agents are an unusual British institution, which has existed for more than two hundred years. They originally dealt with supplying British colonies with all their material requirements. They also printed the colonies' stamps, minted their currencies, held their monetary reserve and performed many other economic functions for the colonies.

Allen Challis left the Crown Agents in November 1973 to become the deputy chairman of a very successful bank in London, called First National Finance Corporation (FNFC). FNFC was the second largest secondary bank in England. Only the big four banks in England,

Barclays, Westminster, Lloyds and Midland, are called primary banks. All the others are called secondary banks. FNFC had grown tremendously during the preceding fifteen years in financing real estate development projects. In late 1973, when the real estate market in England started to weaken, FNFC came to the conclusion that they had to diversify. They chose energy development as their next target. For that purpose they hired Allen Challis, who had vast international experience, gained at the Crown Agents. Challis, in turn, hired a young American lawyer, Bill Kane, who lived in England and was familiar with the oil industry. The two of them were put in charge of developing FNFC's involvement in energy affairs.

At long last we were dealing with willing and able buyers.

As I had expected, the price of fifteen million dollars for the company's assets, actually sixteen and a half (which included repayment of one and a half million dollars to the Israeli Government) was not an obstacle.

The mix of our holdings, which included large areas in Africa and an economic interest in the North Sea block, looked very attractive to them. This was the foundation on which they could build and create a large oil company. The fact that they would have to supply oil to the State of Israel from any new discoveries did not deter them. On the contrary, they viewed it as a plus to have a ready market for their future production.

The negotiations took considerable time, one reason being that, in buying a foreign company, they had to review thousands of documents. Still, matters proceeded smoothly with no big obstacles arising.

The fact that Burmah Oil was a 10 percent shareholder in PEDCO became a very advantageous and beneficial factor. For the bank to be partners with Burmah Oil meant the height of success and acceptance into British industrial society. Burmah Oil had been in existence for over one hundred years. They discovered oil in Burma in the nineteenth century. Burmah Oil created the mighty British Petroleum Company and still owned (in 1973) 23 percent of BP shares. They were the largest shareholder in BP after the British Government. The participation of Burmah Oil in the ownership of PEDCO gave FNFC the confidence that they were taking the right step. It also helped them considerably with the Bank of England, in gaining the necessary approval for the purchase of our company.

I remember having a conversation with Jacob Salman during the period of negotiations with FNFC. Salman was the head of the Government Companies Authority and a member of our Board of Directors. He said: "You are selling the company. What will happen to you?" I replied that there was a remote possibility that the bank would ask me to continue managing the company, as they did not have any oil expertise. "If they ask me, I will think about it. If they do not ask, it is fine as well." Zevi Dinstein, the czar of the Israeli oil industry, had mentioned to me that if we did sell the company and I were to be free, he would like me to be in charge of the whole oil exploration sector of the country. I even had dreams of starting afresh, after the sale, and creating a new I.N.O.C.-PEDCO. Therefore my personal future was of no great concern to me.

Finally, sometime at the beginning of March 1974, we started writing the sale and purchase contract. It was the "winter of discontent" period in England. A long and painful coal miners' strike was on. Electricity was available only three days a week. The typists at the bank had to use candles near their typewriters for light.

The negotiations and the writing of the contract were conducted partly in the Signal-Burmah offices in Houston, where Bill Thompson provided us with all legal, secretarial and other facilities. The Signal-Burmah connection made many things easier to accomplish and conclude.

It was quite a complicated contract, involving the Government of Israel, I.N.O.C. and PEDCO. I therefore suggested employing an additional lawyer, to represent the interests of the Israeli Government per se. Adi Ephrat would continue to represent I.N.O.C.'s and PEDCO's interests. We engaged the services of a prominent Tel Aviv lawyer, Alex Alexandroni, the former Israeli economic minister in the United States, to look after Israeli government interests. He followed the negotiations and the signing of the final agreements to their successful conclusion.

When the final version of the contract was presented by Bill Kane, the American lawyer representing FNFC, it included a clause stipulating that I should continue to manage the company, obviously subject to my agreement. Adi Ephrat deleted this clause from every subsequent draft. Bill Kane became very suspicious. Adi explained to him that Israel was a very small country and rumors would start flying that I had arranged this sale to get myself a nice job abroad. FNFC must find another way of dealing with the approval of my employment.

I had been approached some time earlier by both Challis and Kane to discuss the continuation of my service under the new ownership. I replied that, in principle, I had no objections, subject to an adequate exploration budget and the ability to take my senior personnel with me, but that I was not ready to discuss this subject until the contract between all parties was concluded, in order to prevent any possibility of a conflict of interests.

I must admit that it was quite a tempting offer. To continue to manage the company which I had created, to be relieved of Israeli Government involvement, which was such a handicap in the oil world, and to have a twenty million dollar budget for exploration, which was part of the contract with FNFC, was very different from what I was used to. I reported on this initial approach to the chairman of our Board of Directors and to Zevi Dinstein. Both of them were impressed with the offer, but we did not discuss it any further.

There was another important consideration in favor of myself and my senior staff remaining in the Company. It was to safeguard Israel's interests regarding future oil supplies. The other advantage was that Israelis would be involved in an international oil exploration company. They would act as Israel's eyes and ears in this important field. Dr. Dinstein expressed this sentiment in his letter to me which is quoted below.

Finally, a compromise was reached between Bill Kane and Adi, regarding the clause in the contract concerning my engagement by the new Company. The clause was eliminated from the contract. Instead, FNFC wrote a letter, addressed to Dr. Dinstein and the Israeli Government, which stated that they were giving the final draft of the contract to Adi Ephrat as a trustee. This document could be released only after the Israeli Government's approval of Zvi Alexander's move to England and his continuing service with the company. If, for any reason, this approval would not be forthcoming, the contract should be considered null and void and not released to the Israeli government. In its letter, the bank added that it did not have any expertise in the oil exploration field; therefore, the purchase of the company without its management was a non-starter for it.

I later found out that the continuation of my services was also a condition made by the Bank of England. It was so stated in the Bank of England's letter of approval for the transaction. The Bank of England was seriously concerned about the lack of expertise of FNFC in the oil sector. Allen Challis disclosed this fact in a letter to Dinstein in late 1974. He wrote that they had had lengthy discussions with the Bank of England regarding my employment. The Bank of England was concerned that I might refuse to join and therefore asked for FNFC's assurances in this regard.

The Government of Israel, in a Cabinet meeting, approved the transaction and FNFC's condition that I move to England. When we finally transferred to London at the end of July, I received two official letters, copies of which follow below.

FIRST NATIONAL FINANCE CORPORATION LIMITED

BANKERS

P.O. BOX 505
FIRST NATIONAL HOUSE
FINSBURY PAVEMENT
LONDON EC2P 2HJ

01-638 2855
TELEX 887518
CABLES FIRNAT LONDON EC2

REGISTERED OFFICE FIRST NATIONAL HOUSE FINSBURY PAVEMENT LONDON EC2P 2HJ   REGISTERED NUMBER 59614 ENGLAND

Dr. Z. Dinstein, Esq.,
Director of Petroleum Affairs,
Ministry of Finance,
Jerusalem.   March, 1974,

Dear Sir,

Re OIL AND MINERALS LIMITED ("OMCO")
PETROLEUM EXPLORATION AND DEVELOPMENTS CO. LTD. ("PEDCO")

1. We refer to our letter of the 19th February 1974, addressed
   to the Israel National Oil Co. Ltd. ("INOC") concerning
   the Agreement for the sale and purchase of INOC's shares
   in OMCO and PEDCO (the "Agreement") and your letter to INOC
   of 28th February 1974, which forms a reply thereto.

2. As indicated by us throughout the negotiations (which indication was passed on to you), our Group does not have availab
   to it a management team sufficiently conversant with the oil
   business which can take over the management of PEDCO and
   OMCO, and it was our expressed intention to purchase INOC's
   interests in PEDCO/OMCO provided the management of the said
   Companies will be free, should they so wish, to be employed
   by our organisation. Accordingly, it is a pre-condition
   on our part (although not stated as such in the Agreement)
   that neither INOC nor the Israel Government will prevent
   the present management team of INOC from taking office with
   OMCO/PEDCO under their new ownership.

3. It is our intention to offer to Messrs. Alexander, Eliezri,
   Aharoni, Eizen, Ballot and possibly another one or two
   geologists presently in the employ of INOC to take office
   with PEDCO/OMCO immediately following Completion of the
   transaction and we would request your assurances that neither
   INOC nor the Government of Israel will raise any objection
   to same.

4. To avoid any possible conflict of interests we have not so far, except for expressing our general need and intention as above, either offered to or discussed with any of the individuals concerned any terms concerning their future employment under our organisation and such employment will, needless to say, depend on us negotiating terms with each of the persons concerned.

5. Counsel for INOC has indicated to us that in discussing this matter with you prior to the issuance of your above letter, you have raised no objection to our foregoing demand.

6. Accordingly, so as not to unduly delay Completion of the transaction, we have today caused the Agreement and various additional documents pertaining thereto to be executed on the assumption and condition that no objection will be posed by either your Government or INOC to the future employment of the above individuals in our organisation.

Should such objection be raised, kindly consider the above Agreement and documents as not having been executed and do not submit the transaction for the approval of your Government.

Yours faithfully,

FIRST NATIONAL FINANCE CORPORATION LTD.

Agreed:

Dr. Z. Dinstein

8.5.1974

Government of Israel
Minister of Finance

Mr. Zvi Alexander
Managing Director, Israel National Oil Company
Maya House
Petah Tikvah Road 74–76
Tel Aviv

Dear Mr. Alexander,

It is a great pleasure and obligation for me to praise your great contribution in developing Israel's oil exploration projects abroad. In spite of the great difficulties that you and your colleagues have experienced, encountering opposition to the activities of an Israeli company in this sensitive field, your devotion to the cause, your perseverance and your loyalty have succeeded in transforming the Israel National Oil Company into a successful and valuable asset. Due to unforeseen circumstances you have succeeded in selling the Company for a more than fair price, while preserving the entitlement for special additional rights.

With your nomination as president of the foreign companies abroad, I am confident that you will continue to contribute your efforts to advance the vital interests of the State of Israel.

Sincerely Yours
Signed: Pinhas Sapir

שר האוצר

ירושלים ט"ז באייר תשל"ד
8 במאי 1974

מס׳

לכבוד
מר צבי אלכסנדר
מנכ"ל חברת הנפט הלאומית
דרך פתח תקוה 74—76
בית מעיא
תל–אביב

מר אלכסנדר הנכבד,

הנני רואה חובה נעימה לציין את תרומתך הרבה בפתרון נושא חפושי
הנפט של ישראל בחו"ל. למרות הקשיים המרובים שעמדה אתה ועובדי
החברה בהתנכרות לפעילותה של חברה ישראלית בתחום רגיש זה ידעת
בדבקותך למטרה ובנאמנותך להביא את חברת הנפט הלאומית לחברה
בעלת ערך ולמכרה עקב התנאים המיוחדים במחיר נאה תוך שמירה על
הזכות לקבלת זכויות נוספות.

עם התמנותך למנכ"ל החברות הזרות בחו"ל בטוחני כי תמשיך ותוסיף
לתרום לקידום ענייניה החיוניים של המדינה.

בכבוד רב,

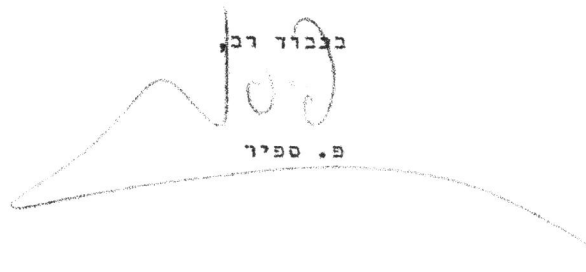

פ. ספיר

28.8.1974

Government of Israel Government Advisor on Oil and Energy
Supervisor of Fuel Affairs

Secret

Mr. Zvi Alexander
Tel Aviv

Dear Sir,

I have been informed that Oil and Minerals Company have made you an offer to join their senior management team abroad.

In the light of the world-wide developments in the field of oil, we take a favorable view of the involvement of Israeli personnel in international oil companies, and your joining of the above mentioned oil company in that context is regarded as a mission which is associated with the State's interest and we fully support it.

I wish you success in your new assignment.

Sincerely Yours
Signed: Dr Z Dinstein

מדינת ישראל

28.8.1974
משרד האוצר ירושלים
מס

היועץ לממשלה בעניני נפט ואנרגיה
הממונה על עניני הדלק

ס ו ד י

לכבוד
מר צבי אלכסנדר
ת ל - א ב י ב

א.נ.,

ידוע לי כי הוצע לך ע"י חברת אויל אנד מינרלס לטד.
להצטרף לצוות הבכיר של החברה בחו"ל.

לאור ההתפתחויות בנושא הנפט בעולם, אנו רואים עבורת אנשי
נפט ישראליים בחברות נפט בינלאומיות, ובמסגרת זו את הצטרפותך לחברה
הנ"ל כשליחות הקשורה באינטרס המדינה, ומחייבים אותה.

באיחולי הצלחה בתפקידך החדש.

בכבוד רב,

ד"ר צ. דינשטיין

Chapter Twenty-four

The Vietnam Disaster

We signed the contract with FNFC in Houston on March 25, 1974.

That night, Minister Sapir called Alex Alexandroni, the lawyer who represented the government's interests, asking whether FNFC had already deposited the first installment into the Israeli Government's account. As this was several days before the beginning of a new budget year, it was important for Sapir to have the millions of dollars of FNFC on the government's books prior to April 1. It would improve the government's foreign currency balance. Alexandroni assured him that the money would be wired the following day.

We had another very important matter to solve prior to the conclusion of the contract. This was the tax liability of the Israeli Government to the British Inland Revenue. The liability arose from the fact that the North Sea economic interests became the dominant asset in the package which the FNFC purchased. There was a discovery on our block in the North Sea, while our Ghanaian holes had turned out to be dry. The balance therefore changed. The value of the African holdings was lowered to five million dollars, while the value of the economic interest in the North Sea rose to ten million dollars. The tax liability was four million dollars.

We had an excellent tax advisor in London. He came up with a brilliant idea which eliminated this tax liability, namely to create a situation whereby the North Sea asset would constitute less than 50 percent of the company's assets. In order to achieve such a situation it was necessary to inject, from abroad, enough money which, together with the African assets, would be a sum larger than the North Sea interests, on the day of the closing. Accordingly, twenty-four hours before the closing, we borrowed, from the Israeli treasury in New York, enough money to create such a situation. The money was returned to the Israeli treasury in New York the following day. But on the determining day, March 25, the North Sea assets did not constitute a majority of the holdings. Our tax consultant received an official letter from the Inland Revenue in London confirming that, in view of that situation on March 25, there was no tax liability.

We had our first Board meeting in the Signal offices in Houston one day after we signed the contract. Bill Kane represented the bank, Bill Thompson represented Burmah Oil, and I the company. We had various legal and organizational decisions to make and to prepare minutes. We also had to review the status of all our licenses and record that status in the minutes. At the end of the meeting, I suggested that the next Board meeting take place in Tel Aviv, on May 1, 1974. This proposal was enthusiastically adopted, especially as I extended an invitation also to the wives of the directors.

* * *

The meeting in Tel Aviv was a great success. Allen Challis, Bill Kane and his wife, Bill Thompson and his wife, Byrdie, were the participants from abroad. We hired a special plane which took us for a trip to Saint Catherine's Monastery on Mount Sinai and to Sharam El Sheikh at the southern tip of the Sinai peninsula.

We had a big reception at the Hilton Hotel in Tel Aviv, which everybody who was anybody on the political and economic scene of Israel attended. Ministers, industrialists, bankers—all came to meet the foreign guests. They all wanted to see the foreign bankers who paid millions of dollars for some "funny" concessions that most of them did not know even existed. During that party, Emanuel Racine, the president of Delek Oil Marketing Company and the doyen of Israel's oil industry, approached my wife and said: "Congratulations, Rachel, but I want you to know that your husband will never be forgiven." My wife wondered: "Mr. Racine, why are you saying this?" He replied: "A government company by nature has to lose money. Your husband has shown that a government company can make millions of dollars of profit. This is unforgivable." The funny thing was that he was half serious.

During the Board meeting in Tel Aviv, the subject of exploration in South Vietnam came up. South Vietnam opened its offshore for oil exploration in late 1972. As the offshore of Vietnam includes the Mekong river delta, it was considered to be a very attractive area for oil and gas exploration. Most of the large river deltas in the world are very prolific. Sediments rich in organic material were transported by those rivers to the sea for millions of years. This organic material became the source for large oil and gas accumulations.

* * *

We first approached the South Vietnamese Government in early 1973 but we were refused because of our small size. When we felt that the Bank was going to buy the company, we contacted the government in Saigon again and they welcomed us.

Allen Challis and Bill Kane asked Bill Thompson for his views regarding South Vietnam. Thompson recommended it highly. The Board therefore approved the venture. The question was how many blocks we should apply for. We started with one block, but one Board member said: "Well, if we apply for one, we might get nothing. Let us apply for three blocks." The mood of the Board was bullish, the weather was nice and the country was beautiful, so that by the end of the meeting the decision was reached to apply for five blocks in the hope of obtaining one block.

For each block, an applicant had to deposit one hundred thousand dollars earnest money. That sum was going to be applied as part of the signature bonus, when the licenses were to be issued.

Each company also had to offer a signature bonus for each block for which it applied. The highest bidder would be allocated the relevant block.

Following the Board meeting, we sent our head of exploration and our lawyer, Micki Deouell, to South Vietnam to choose the five blocks.

They chose the blocks and submitted the applications with a bid of four hundred thousand to six hundred thousand dollars per block, as signature bonus. The South Vietnamese acted in extreme haste, and the awards were made within a few weeks.

We also concluded an agreement to join forces in South Vietnam with Shaul Eisenberg, who had major industrial involvements there. Eisenberg was convinced that South Vietnam was here to stay, that the war against the Vietcong was almost won, and that there was nothing to be concerned about. He, on his part, was putting up factories in South Vietnam and had great support for his activities from the industrial world, especially from America.

* * *

Eisenberg at the time was probably the wealthiest man in Israel. His fortune was considered to be in excess of a billion dollars. He had a very interesting history. As a young man, as a refugee from Germany, he escaped to Japan, where he spent the war years. He married a Japanese girl who later converted to Judaism. During the American occupation, he became a supplier to the American Army and made his first fortune. He later extended his activities into South Korea, Vietnam and China. He bought a Boeing 707 which he equipped for his private use. He had offices in many places around the world, and was a one-man conglomerate. In the 1960s he approached the Israeli Government offering to locate his headquarters in Tel Aviv, on condition that he would not be subject to taxes on any of his foreign activities. If and when he would do any business in Israel proper, he would pay tax on profits derived from activities in Israel.

The Government agreed, and a special law was passed in the Israeli parliament, known as the Eisenberg Law, exempting any and all of his foreign activities from any Israeli taxes and regulations.

* * *

My wife and I went to Ashkelon in late May 1974 for a short vacation. There were no phones in the hotel rooms, as yet, and I went to the hotel lobby to phone our office in Tel Aviv. I was told that we had just received a notification that we were allotted four blocks out of the five we had applied for in South Vietnam. I went back to my room and told my wife to start packing, as we had to go back to Tel Aviv, right away. I told her that something very wrong must have happened. The fact that we received four blocks, out of the total of twenty that South Vietnam was awarding, frightened me. There was fierce competition in South Vietnam. All the major oil companies were there. They must have offered signature bonuses of millions of dollars per block. Had we received one block only, it would have made sense. But all four—that looked like a calamity. I suspected that we might have chosen the wrong areas that nobody else wanted.

The Vietnamese government invited us to come to Saigon, within a month, for the signing ceremony. We had to bring with us the signature bonus payments which came, in our case, to approximately one and a half million dollars plus a twenty million dollar bank guarantee. The bank guarantee of five million dollars per block was to assure the government of the promised exploration investment. It looked like a disaster of major proportions.

We therefore had to find partners immediately who would join us in Vietnam and carry most of the financial burden. We had the list of all the applicants in Vietnam who had not received an allocation. We also did not have the Israeli Government connection anymore, which might have been a stumbling block. Any company could deal freely with us.

I went immediately to the States with our chief geologist, Efraim Aharoni, to try and find partners. We went from one company to another but did not receive any favorable response. Towards the end of the trip, I talked to Forest Molsbery, the chief geologist of Hamilton Brothers, who were our partners in Ghana. I asked him why I was receiving such a negative response. Forest said: "Zvi, I must level with you. Who chose these blocks? There is basement rock sticking out in the middle of your blocks. There is an article in a recent AAPG journal (American Association of Petroleum Geologists) describing the offshore of South Vietnam which shows the basement areas prominently."

A seismic survey of the offshore of South Vietnam was also available. It was a crime that we had not known about it, and did not buy it in advance of the bidding. Reviewing the seismic data, which we purchased thereafter for thirty thousand dollars, would have shown us the basement areas immediately.

This was a real disaster. Oil accumulates only in sedimentary rock, which was deposited in the sea over millions of years on top of the basic rock, called "the basement." As our blocks were mostly basement, there was no hope of finding any oil there. No hope in hell to find a partner.

I went to Houston to see Bill Thompson. It was Saturday. We drove to the office and picked up the AAPG journal and, sure enough, it was "spelled out in spades."

What do we do now? What a shame! What shall we tell the bank? Bill Thompson felt very bad. After all, he had recommended the project. His company was supposed to give us technical advice. Although we did not consult them on the Vietnam application, he still felt kind of responsible.

Upon my return from the States, I asked Micki Deouell, our concession lawyer, to go to Saigon. I promised to follow him the next day. Micki met me at the Saigon airport, white as a sheet. He said: "We will be lucky if both of us don't end up in jail."

He told me that the Vietnamese cabinet had met the previous night, after hearing that we were not going to sign for the relevant blocks and the payment of the signature bonuses would not be forthcoming. In addition to the loss of anticipated income, it was a tremendous loss of face to the Vietnamese minister of mines and his director general. Loss of face in the Far East is one of the biggest crimes, and we had put them in this untenable position.

The foreign press in Saigon got wind of our predicament. They continually phoned us at the hotel asking for a statement.

I tried to explain to the Vietnamese government that it was an honest mistake and that they should not expect us to spend money on hopeless exploration. I suggested that the government release some other blocks to us, from the national reserve which they kept for themselves. They did not agree. Finally, after two days of discussions, they believed us and let us go back. The earnest money which we had deposited to the tune of five hundred thousand dollars was lost. We had lost not only face but half a million dollars as well.

When the plane took off from Saigon airport, Micki and I finally believed that we were not going to spend the next few years in a Saigon jail.

Six months later the story was over. Saigon fell to the Vietcong forces.

Chapter Twenty-five

Moving the Company to London

We started thinking about the move to England. It was a difficult period. We had two very distinct groups of people in the company. Those who were going to England were happy, busy reorganizing their lives, their wives resigning from their jobs, renting out their houses, etc.

The other group, those who were going to remain in Israel, with their future uncertain, were obviously frustrated and unhappy.

The Government did not nominate a new head for I.N.O.C., and I was in the strange position of being both the head of I.N.O.C. and the head of the foreign company, PEDCO, which had purchased I.N.O.C.'s assets. In these two capacities, I was writing checks from one company to the other, being the paying and the receiving party at the same time. It was really a bizarre situation.

When I complained to Dr. Dinstein about it, he said that he trusted me to fulfill both jobs honestly. Finally, sometime in late June I resigned from the Board of I.N.O.C. and from being its managing director.

Still, I had to continue to act for both sides until we finally left for England at the beginning of August 1974.

One serious mistake I made during that period was not to deal with my pension entitlement. Jacob Salman, the head of the Government Companies Authority, offered me a full pension, i.e., 70 percent of my most recent salary. I refused to take it, and for the life of me I cannot understand why. I would have had a substantial pension for life, linked to the salary of the president of the Haifa Refinery. In recent years, the salaries of Israeli executives became quite comparable to those of their American peers. I could have been collecting these large sums for close to thirty years.

I am probably one of the very few Israelis, who served in the British Army, as well as in the Israeli Army, and in high government positions, who is not receiving any pension at all. Almost all of my friends receive several pensions: An army pension for their service until the age of forty, and one or more pensions for their service in various government or other institutions, or companies, after leaving the army. Somehow or other, I am the exception. This is ironic in view of the fact that, in addition to my long service, I also contributed sixteen and a half million dollars to the Israeli treasury. I know of no other Israeli who achieved such a feat, or even a fraction of it.

Instead of dealing with my entitlements and pension, I spent a lot of time trying to get approval for the bill our attorneys submitted for the PEDCO transaction. It was quite a substantial amount. In my view they were entitled to it. I divided the sum in two, in order to make it more palatable in Israel, with the intention that 50 percent of the bill be settled later by our new foreign company. The Bank was rather perplexed as to why we were paying the Israeli attorney's fees. They agreed to it when I told them that I had promised this to be so. They accepted and honored my undertaking.

As to the other 50 percent, that I.N.O.C. had to pay, I had a lot of lobbying to do to have it approved. Dinstein refused to approve it, claiming that the sum was excessive. I had to speak to the Minister of Justice, Haim Zadok, to have it finally settled.

I remember sitting one day with Adi Ephrat and his partner, Ezra Goddard, discussing their bill. They pointed out that now, when I was leaving the country, the subject of oil exploration, both in Israel and abroad, that had taken up a large part of their firm's time and was one of its main occupations during the last ten years, was finished and over. Therefore, their bill also reflected that fact.

How wrong they were! Twenty-five years later, their firm was still dealing with oil exploration matters—first with the Sinai Desert production before it was returned to Egypt in 1981, and thereafter with the various public issues devoted to raising money for exploration in Israel.

Not that any oil has been found in Israel in the last forty years (except for the deep offshore gas discoveries in 1999, as described below). But hope springs eternal and the Israeli public, like that of many other countries, loves to gamble. Therefore there is a lot of room for various entrepreneurs, lawyers, underwriters and, obviously, geologists to get into the act.

I often wondered if, had I stayed in Israel, I would still be involved in this futile exercise. I do not think so. To look for oil where God apparently did not destine it to be, and to take trusting people's money for such a hopeless undertaking, goes against my nature.

After we moved to England, an Israeli geologist asked me to help him get work in the North Sea. I arranged this and he worked for a year as a well-site geologist on a drilling platform in the North Sea. He occasionally joined us for dinner in our home in London. During his last visit he complained that he could not stay on the offshore platform any longer. His main problem, he said, was that he could not drink liquor. All the other employees on the rig floor started to drink after the end of their shift, whereas he was unable to join them as his stomach could not take it. What should he do, he asked.

I told him that there were two schools of thought in Israel—one believed that oil can be found there and the other group disagreed. If he belonged to the first group, he should obviously go back home; if he belonged to the second group, he had a serious problem.

He asked me who does believe that oil can still be found in Israel. I was taken aback. I said: "What about Dr. Kashai?" Dr. Kashai was the doyen of Israeli geologists, who had received a Ph.D. degree in geology in Hungary. My guest replied: "In what language did you discuss this subject with him? Ask him in Hungarian and he will tell you what he really thinks."

My geologist guest returned to Israel and he is still involved in oil exploration there. As I said before, "hope springs eternal," and one also has to earn a living—which is a very strong motive.

* * *

In 1999, a commercial gas discovery, in deep waters opposite Ashkelon, off-shore Israel, was announced by a consortium of companies headed by Samedan Oil Company from Oklahoma. Following this discovery, more gas was found by Samedan and by Isramco, an Israeli company. Both companies reported that their combined proven reserves are a trillion cubic feet of gas, a very respectable quantity with a market value, over time, of three billion dollars. This quantity could satisfy the Israel Electric company's needs for five to seven years.

The technique of drilling and producing in deep waters did not exist until the end of the 1980s. This technique was developed in the last ten years by American oil companies drilling in the Gulf of Mexico, in the waters of Texas and Louisiana. From there, the technique was copied and appilied in other parts of the world. However, in the period of oil exploration described in my book, it was impossible to think of exploration, and especially production, in deep ocean waters.

I believe that all the dry holes drilled in Israel at a cost of hundreds of millions of dollars, coming mostly from the pockets of Israel's taxpayers during the last forty-five years, did not contribute at all to the discovery of gas in the deep waters offshore Israel. These gas discoveries would, most likely, have been made anyway by foreign oil companies which discovered tremendous quantities of gas offshore Egypt in the Nile delta. These companies, as a direct continuation of their discoveries in Egypt, would have come to explore offshore Israel. In fact, British Gas, one of the companies that made the discoveries in Egypt's Nile delta, received exploration rights from the Palestinian Authority offshore Gaza and has recently discovered a very large gas deposit there. British Gas also joined the Isramco Group offshore Israel.

The Egyptian gas originates from the sands, full of organic sediments, which the Nile deposited in its delta over the past millions of years. It is quite possible that those sands travelled further northwards, and were deposited offshore Gaza and Israel.

This theory, of finding oil in the deep waters offshore Israel, was conceived some forty years ago by Dr. David Niv, nicknamed "Bibi," who was then Israel's petroleum commissioner. He was considered a dreamer, especially in view of the fact that there were no technical means, at the time, to drill and produce in those deep waters.

Dr. Eli Rosenberg, former chief geologist of Nafta, who was responsible for many of the wells drilled in Israel in the last forty-five years, conceived also the idea of drilling in deep waters offshore Israel. He founded the Avner company for this purpose, and then found the American Samedan company which joined him. They drilled the first well in the deep offshore of Israel opposite the city of Ashkelon, where the first discovery of commercial gas was made. Samedan, which heads the group, owns a 40 percent interest in the project. The other partners are Avner, Delek, the oil marketing company, and Reading & Bates, a drilling contractor from the US.

In 1997, Dr. Rosenberg asked me to help him find a foreign company to join this project. I tried to assist him but was not successful. Only recently have the big discoveries in the Nile delta in Egypt come to the attention of the world oil community, but in 1997 it was very difficult to arouse interest in such a project in Israel. Dr. Rosenberg deserves all the credit for his perseverance in searching without respite for a serious partner. He found Samedan, a well-respected oil company, and more power to him.

Some people are not happy that a foreign oil company will profit from a discovery on our shores. I can assure those doubters that, as far as the State of Israel is concerned, it hardly matters whether the oil and gas were discovered and produced by a foreign or a local company. In every instance, the state receives approximately 75 percent of the total income from production, through royalties and various taxes—without the state having to invest any funds in exploration and production. Furthermore, only the state has the right to decide whether the discovered oil and gas will be directed to the local market or exported, and on any division between these two markets.

The discoveries of the gas offshore Ashkelon brought another important benefit. Egypt, which understands that due to those discoveries Israel's dependence on imported fuel will greatly diminish, is very eager to sign a long-term contract to supply gas by pipeline to Israel. I had thought about such a project more than twenty years ago, as described in chapter 30.

Chapter Twenty-six

The "Winter of Discontent" in England

We arrived in England at the beginning of August, 1974. The day we arrived we realized that this was a different country from the one we had previously known, and a different bank from the one that we had made the agreement with.

England was in deep financial crisis and FNFC was tottering on the brink of insolvency.

We realized that, had we not signed the agreement in March, the transaction would have never been consummated. Had the negotiations continued for another two weeks, FNFC would have had to withdraw from the deal.

At the beginning of 1974, FNFC was still flying high. When the first large bank collapsed in London, the Bank of England asked FNFC to save it. FNFC made the condition that the Bank of England participate in the rescue operation, which it did. This was the first time in history that the Bank of England participated with any private bank in a commercial operation.

During April 1974, FNFC's situation started to deteriorate. In the same month, the collapse of the real estate market began. This created a chain reaction of bankruptcies—first the real estate companies, and then the banks which had lent them money.

The situation was so serious that there was fear of another "1929" depression which would ruin the British economy for many years to come. The Bank of England was forced to come to the rescue. It created, together with the four big banks—Barclays, Westminster, Lloyds and Midland—a "Life Boat Committee" in which they put together five billion pounds. (This, at the time, was close to ten billion dollars). It was a tremendous sum, equalling close to one hundred billion dollars in today's money. The purpose of the Life Boat Committee was to rescue those banks and other financial institutions which were worth keeping afloat. The Life Boat Committee most probably saved England from entering another 1929-type depression.

* * *

FNFC prepared an office for us in the West End of London, one of the most prestigious and expensive areas of England's capital. There was one very large room in the office and six or seven small rooms. Bill Kane, who became the chairman of PEDCO, took the large room, while the smaller rooms were assigned to the rest of us.

Five of us had come from Israel: the head of exploration, the chief geologist, the chief geophysicist, our financial man and me. In addition, the bank furnished us with a British lawyer and a British accountant, who were both going to serve our company full time.

There was very little to do. The big hopes of a twenty million dollar exploration budget, that the bank had undertaken to furnish us, disappeared. Slowly but surely, we got the feeling that the bank wished never to have met us. The bank had a six hundred million pound problem; therefore, their investment of ten million pounds (sixteen and a half million dollars) in purchasing our company was an insignificant sum. It was only 1.5 percent of their problem. This bizarre situation went from bad to worse in the course of time.

* * *

I had several amusing encounters during this initial period, which taught me how different we Israelis are from the British.

The first such experience was related to bookcases. The bank had rented for me a large apartment on Park Lane, one of the nicest locations in London. It was two minutes away from our offices. We had shipped several crates from Israel containing our books, but there was no place to put them. There were no bookcases in our apartment. I started negotiating with the porter, then with the manager of the building, and then with the management company, but to no avail. I remember a conversation I had with a man named Cross from the management company. I finally said to him. "Mr. Cross, do you have any books at home?"

"Yes," replied the man

"Where do you keep them?"

"In a bookcase."

"This is what I have been asking for the whole last month!"

While all these "negotiations" were going on, I was walking one evening on Mount Street, near our apartment, when I saw a furniture shop which had a large arrangement of shelves in the window. These shelves could solve our "bookcase problem." I asked the salesman: "How much is this bookcase?"

He replied: "Sir, is this for a home or for an office?"

"It is for my apartment."

"I am very sorry, Sir, we sell only to offices."

"Fine, our offices are nearby. I will buy it for our office."

"We only sell made to measure furniture."

"Fine. But this size is exactly what we need, therefore you can save the cost of measuring."

"I am sorry, I cannot help you."

"Why not?"

"You said in the beginning that you wanted it for an apartment and we do not sell to apartments."

Such an encounter could not happen in a thousand years in an Israeli shop. It would not happen in America, either.

The second experience was when our British accountant came to my room asking me to sign a check for two hundred and forty pounds made out to an air-conditioning company. I asked him, "Charlie, what is this?" He replied: "We received the bill."

I said: "Charlie, for God's sake, this is a newly reconditioned office and we haven't been here more than a month. How come this bill? Maybe it was intended for the building next door. Call the company up and find out."

He came back a day later saying that it was a mistake, the bill should have been for forty pounds instead of two hundred and forty. It was for a filter replacement. I did not think that we should be charged for the filter either, but I did not want to argue any more and signed the check.

A couple of months later I was discussing something with Jonathan North, our British internal lawyer, when he suddenly said: "For you Israelis a bill is a subject for discussion."

"Jonathan, are you insane? Did you intend us to pay a bill for repair of air-conditioning which was not needed, and probably not even done?"

"No, no," he replied, "but I am making a general observation."

Still the British ruled the largest empire on earth for hundreds of years. So maybe there is something in the way their minds work.

* * *

Bill Kane, the American lawyer and PEDCO's chairman, was very unhappy. He was hoping to be a big oil executive, with an empire encompassing oil, gas and coal; suddenly there was very little left of all these dreams. He would probably have settled for managing only PEDCO, but here was I standing in his way. Had he been smart enough, he would have gone to the directors of FNFC and stated his case. He should have said that he was much younger than me, that he was an American, non-Jewish and a lawyer, and that he would like to take my position. It is possible that he could have convinced the bank to do that.

Instead, he started floating accusations that we were financially irresponsible. He presented his accusations to PEDCO's directors during a Board meeting in London, in mid-December 1974; the accusations were reviewed and proven to be false. Bill Kane was fired on the spot. Within one week he was out of England and out of our sight.

Ananda Krishnan, the Malaysian businessman who introduced us to FNFC, was also involved in our affairs. A Far Eastern group organized by him, together with the Crown Agents, had participated in the purchase of our company. Out of the 90 percent of PEDCO that FNFC bought from the Government of Israel, the bank bought 72 percent and Ananda's group and the Crown Agents the remaining 18 percent.

The crisis in Britain affected Ananda very badly. He tried to convince me that we should use the one and a half million dollars that we had in our account to buy a coal mine. I tried to reason with him that this was not in our mandate, that we did not know anything about coal, and that it was a crazy idea altogether. But people in desperation do many silly things. Apparently Ananda was in that state at the time. As I have mentioned before, he is now a multi-billionaire in Malaysia.

In November 1974, we were moved to FNFC's building on Finsbury Pavement in the City of London. Spending our days in the bank's building was a traumatic experience. Each day you saw fewer people in the bank. Sometimes it resembled a graveyard.

There was very little left for us to do. I read *The Times, The Financial Times, Time* magazine and other publications. By ten or eleven o'clock I had finished my reading material. Once, when I complained to one of the directors of being underemployed, he said: "Why do you bother us? Go skiing, or whatever else you want to do. We have much bigger problems than yours. Our problem is to the tune of six hundred million pounds."

* * *

The Life Boat Committee nominated the deputy head of Westminster Bank to be in charge of FNFC, while Pat Mathews, the former chairman, left.

When I think back on this period, I come to the conclusion that I must have been paralyzed. The shock of the bank's collapse and of the uncertainty of our future must have brought me to a kind of inertia. Instead of travelling and enjoying myself, I sat frozen in the office.

The following example will show how incapacitated I was. In July 1974, during one of our visits to London before moving there, my wife and I stopped off in Holland to visit the daughter of a famous geologist who had worked in Palestine at the beginning of the twentieth century. I was hoping to find letters, or rather envelopes of the letters, he wrote home. I was, and still am, an enthusiastic collector of documents relating to the postal history of Palestine. Although I probably have the best collection on this subject in the world, I have never stopped looking for more old correspondence from Palestine.

We drove to Eindhoven where this lady lived. She received us very graciously and brought from the basement a number of diaries that her father has written in Palestine. I explained to her that my interest lay in postal history. She promised to search through her father's effects. We made a date to see her the next morning. When we woke up, the sun was shining and it was simply a glorious day. My wife convinced me that there would be nothing to see at Ms. Blankenhorn's house and that we should drive back to Amsterdam instead. Nearing Amsterdam, I went into a telephone booth to call Ms. Blankenhorn to apologize for not coming, saying that I had received an urgent call to go back to Amsterdam. She replied: "What a pity. I laid out for you all the letters and postcards which you were so interested in."

The disaster that greeted us on our arrival at the bank in London the following month made me forget this lost opportunity completely, and I did not arrange to visit her later. Such an oversight would never have happened to me if the situation in London had not been so heart-breaking.

Chapter Twenty-seven
Desperate Attempts to Save the Company

In spite of the frozen state of my mind, to which I referred in the previous chapter, I still made several serious attempts to save our company.

The most important attempt was in October 1974, two and a half months after our arrival in England. Uri Lubrani, the Israeli ambassador to Teheran (at that time), visited me in London. Uri and I had become friends when we first met in Addis Ababa, when he was the Israeli ambassador to Ethiopia. When he returned to Israel from Ethiopia, I invited him to become a member of the Board of Directors of I.N.O.C. As a result, he knew everything about our company and the details of the sale to FNFC.

During our long discussions in London, I made the suggestion that the Iranian National Oil Company might buy our company. Uri came up with a brilliant idea: "Why don't the Iranians buy the whole bank, including your company? You will then have all the money in the world."

Upon his return to Iran, he presented this idea to the country's prime minister and to its Minister of Finance. They both liked it.

On December 9, 1974, a meeting took place in Paris. Dr. Fallah, the deputy head of the Iranian National Oil Company, accompanied by Uri Lubrani, arrived from Teheran. Dr. Fallah was one of the most important personalities in Iran. It was obvious that if he came to the meeting, he already had the green light to discuss the purchase of the bank. Allen Challis and Adrian Evans, both of them directors of FNFC, and I came from London. At the end of the meeting, it looked like it was a done deal. We felt that the company and the bank were going to be saved.

Upon Fallah's and Lubrani's return to Teheran, the Iranian Government took stock of Iranian acquisitions during 1974. During that year, they bought 25 percent of Krupp, the largest steel producer in

Germany. They had also made a number of other very large investments. They had placed orders for several billion dollars' worth of arms. The sum total of their expenditures was so vast that the Shah ordered a halt on any additional new purchases. Our project fell into this category.

Had we approached the Iranians one month earlier, the Iranian National Oil Company would probably have become our new owner. By the time Ayatollah Khomeini came to power, in 1979, we probably would have had a very large oil company and all the money we would ever have needed. The Iranian Revolution would probably not have concerned us at all. On the contrary, it might have worked in our favor.

Uri Lubrani is now the man in charge of Lebanese affairs in the Israeli government. He has filled this post successfully for the last fifteen years or more. He served under five or six different Israeli Prime Ministers. The governments change but Uri remains at his post.

* * *

Another possibility that I explored in early 1975 was with Barry Damson. Barry was the controlling shareholder of Damson Oil Company in New York. He came to London and spent a whole week talking to the bank about merging PEDCO with Damson Oil Company. We already agreed that there would be a three-man executive committee running the combined company: Barry Damson, John Lomax, his head of exploration and an old friend of mine, and me. Damson probably got cold feet, upon his return to New York, when he realized that an additional twenty-five to thirty million dollars would be required as our 5 percent participation in the cost of developing the Thistle oil field in the North Sea.

* * *

I explored another merger possibility with Basic Resources Company of New York. Basic Resources was involved in exploration in Guatemala, where they had a small producing oil field. I met with the president of Basic, but he thought that he was already producing oil, while we still needed considerable funds to develop Thistle. Our talks were not successful.

* * *

Throughout all that period, serious talks were held with the Israeli Government, to buy back PEDCO. The Bank offered at one stage a discount of 50 percent on their purchase price. They were ready to lose eight million dollars as they did not have the funds needed for the development of the Thistle oil field. The problem was that the Israeli Government was interested in oil and not in profits, and rightly so. As PEDCO had only an economic interest, and no right to receive the oil production from Thistle, Israel was not interested.

There was one instance when negotiations almost succeeded. The formula worked out was that the bank would buy back from Burmah Oil the right to own directly 5 percent of Thistle's output. The bank would then sell the Thistle interests to the Rothschilds, who would front for the Israeli Government. However, as the bank was by then in almost total disarray, there was nobody to carry this project through.

* * *

During that period, bills started to arrive to cover the cost of the development work that was being conducted in the Thistle oil field. I went to see the banker from Westminster Bank, who had been nominated by the Life Boat Committee to administer FNFC affairs. I told him that, in my view, FNFC should pay those bills. I said: "Look, you are selling off real estate assets for forty cents on the dollar. One of the reasons for the big discount you are giving is your fear that the value of those assets might continue to decline further. In the case of the Thistle development costs, you know for sure that every dollar you invest today will bring several dollars of profit when the field starts producing. There is no doubt about it."

He replied: "I agree with you, Mr. Alexander, but we must slim down." He repeated: "We must slim down. We decided to make no further investments of any kind." Eventually the bank sold the North Sea Thistle interest back to Burmah Oil at a substantial loss.

It is possible that I could have found an appropriate solution for PEDCO had I had the same freedom of action in the bank that I had had in Israel. However, the situation in the bank was completely different. Whereas in Israel I did not have any money, but I could go anywhere and do anything I wanted, in the bank we were foreigners who were viewed with suspicious eyes. Therefore our freedom of movement and of negotiations was very limited.

* * *

One more illustration of the mood of foreboding in England in 1975 and 1976 may be of interest. Adrian Evans, the young FNFC director, told me one day that he had sold his flat in London, which was close to Regent's Park, and had bought a house in St. John's Wood, nearby. When I asked why he did it, he replied that conditions in England were going to be so treacherous and difficult that he must be in control of his expenses. He must be able, if needed, to close several rooms in the house to cut the cost of heating and electricity. This he would not be able to do in an apartment building.

Adrian Evans belonged to a long-established and well-to-do family. He must also have had some savings, as he had been a vice-president of Citibank in New York before returning to London. Still, he had this terrible fear of an approaching calamity in England.

Chapter Twenty-eight

Newspapers Attacks

Due to the serious situation in FNFC, we could not start any new ventures, but we had to continue to operate our remaining valid licenses in Africa. In October 1974 we concluded a farm out agreement with the General American Oil Company, of Dallas, Texas, regarding our five blocks offshore Gabon. General American were also our partners in Ethiopia. According to that agreement, General American had to conduct a seismic survey in our blocks in Gabon in order to establish drilling locations. The survey was very much hampered by a layer of deep salt which covered the sedimentary section in our licensed areas. This salt layer masked the seismic reflections from the underlying strata. The result was an unclear seismic picture. General American came to the conclusion that there was too much risk in drilling features that were not clearly defined. They gave us the result of the seismic survey without demanding any payment and transferred their rights back to us.

* * *

During that period we also had an interesting contact with Exxon, the largest oil company in the world. Exxon had a discovery offshore Ivory Coast, in an area not far from our Tano license in Ghana. The possibility of cooperation with us in Ghana was discussed at an executive committee meeting at Exxon. Due to the economic slump in 1974/75, even Exxon decided to reduce their exposure. Many new projects were not approved. Ours suffered the same fate.

* * *

During the course of the year, military coups took place in Ethiopia and in Madagascar. The coups made our concessions in those countries inoperable.

* * *

In the preceding chapter, I mentioned Guatemala in regard to Basic Resources. I was very intrigued about the oil possibility of that country. In the early 1970s, very large discoveries of oil were made in Mexico, and it looked like the same geological setting might exist in Guatemala.

Guatemala was the only foreign country that Adrian Evans, the director of the bank, allowed me to investigate for new exploration possibilities. I went there in May 1975 and spent a week in Guatemala City. Although the Israeli ambassador to Guatemala, General Yitzhak Pundak was a dear friend of mine, I did not succeed in creating the right contacts, and the trip was not productive.

* * *

During the Christmas Holiday season of 1974 I went home to Israel for a week. One morning Zevi Dinstein woke me up, with an angry telephone call, asking me whether I had seen the *Ma'ariv* newspaper yet.

The paper had a whole page article, featuring my picture, saying that I had returned to Israel to consult with the government about the bank's inability to pay the rest of the purchase price. (The bank had actually paid all the money earlier than it was obligated to.)

The article continued with allegations that we have sold part of the North Sea oil field, which was the only hope for Israel to obtain oil from a safe source, in order that Zvi Alexander would get a nice job abroad.

Other newspapers followed suit and a full blown newspaper campaign started against me and the "stupid government" which allowed me to "commit the crimes."

A geologist who previously worked in I.N.O.C. was quoted in the slandering articles. He stated that he was against the sale of the company to FNFC. He "forgot" that he was the most enthusiastic employee when the sale was agreed upon and the happiest man on earth when he moved with us to England. Once the newspaper "party" started, some of the I.N.O.C. personnel that carried a grudge because they were not chosen to move with us to London joined the "festivities."

The same geologist who initiated the newspaper attacks and who worked with us for several months in London was also the strongest supporter of the diamond exploration venture in South Africa. The diamond venture was under his personal supervision. Now when the attack began it was the most opportune time to mention the diamond "folly." It was attacked with a vengeance. As the word 'diamonds' always arouses strong reactions in the public, it was a very juicy target for the attack.

The Israeli Military Censorship added to the confusion by disallowing the publication of some of the relevant facts from the articles.

All the well-wishers, in the government and elsewhere, who had been so excited and happy about the windfall of sixteen and a half million dollars, and who had congratulated me on every occasion, suddenly disappeared. When the shit hits the fan, everybody tries to escape in order not to be covered by it.

I requested the former Chairman of the Board of Directors of I.N.O.C. to publish a full page advertisement in the newspapers quoting the Board's resolutions to sell the Company. To quote their insistence and constant pressure on me to find a buyer quickly, before the assets may be lost. First he agreed but then he changed his mind. He was afraid to do it.

Jacob Salman, the head of the Government Companies Authority, called and begged me not to publish Minister Sapir's letter to me that warmly praised my achievements and voiced his confidence in my continued service to Israel abroad. The original letter is shown in this book. To this day I do not understand why I agreed to withhold publishing Sapir's letter in the newspapers.

Danny Halperin, the spokesman for the Ministry of Finance sent to me a journalist from *Ma'ariv* to write the true story. The man came to see me but I realized that he was not interested. The only story worth writing about is that a man has bitten a dog. The reverse of such a story, "dog bites man," though true, is of no interest to the public. It does not sell newspapers.

There was only one consideration which might partly explain the shameful and cowardly behavior of the Israeli government. The government may have wanted to protect the secret clauses in the agreements relating to the supply of oil that bypassed the Arab boycott, by keeping mum. Maybe?

* * *

Anybody who was not attacked by a vicious newspaper article cannot understand the feelings of rage and helplessness of the person set upon. You find yourself walking in the street, thinking that the man walking towards you recognizes your face from the picture and has read the articles about you, accusing you of all these "crimes." It is a terrible feeling.

The fact that these articles fell on receptive ears was quite understandable. The public in Israel felt that part of the country's troubles stemmed from its lack of oil resources, while the Arabs drew all their strength from oil. And here was a villain who sold those riches from under them—and the stupid government let him do it.

The fact that we never owned any North Sea oil, and that only due to Signal's generosity, after all its fights with Santa Fe, we had ended up with an economic interest, was not publicly known and very difficult to explain. The story in the paper was clear and simple. According to the story, Israel did have North Sea oil. Zvi Alexander sold it. The crucial question as to how he could have sold something that belonged to the State of Israel, without the government's blessing and permission, was never asked and never answered.

I went to see the Minister of Justice Haim Zadok. I had the impression he felt very uncomfortable, and that he wanted my visit to be as short as possible. On the one hand, he felt ashamed that he and the government did not come out in my defense. He knew very well that this was his government's decision and that he personally had congratulated me on several occasions. He was convinced that this was one of the best economic achievements ever. But on the other hand he did not want to get his hands dirty, he was afraid that he might be attacked as well.

Although almost thirty years have passed since, I still feel the hurt and disappointment at the behavior of my colleagues.

* * *

I have recently read a book about General Ehud Barak, the former Prime Minister of Israel. The name of the book is "The Sixth Medal of Valor." Ehud Barak was the most decorated soldier of the Israeli Army. The number of his daring exploits is legendary. He was the commander of *Sayeret Matkal,* the elite commando unit of the Israeli Army, famous among other daring exploits for the rescue of the hostages of the Air France passengers in Entebbe, Uganda. Barak rose to the rank of Lieutenant General and became the Chief of Staff of the Israeli Army.

During one of the exercises of *Sayeret Matkal*, which was reported by the foreign press to be a dry run for an attempt on the life of Saddam Hussein, the dictator of Iraq, six commando fighters were killed by a missile fired in error. Ehud Barak, the chief of staff at the time of the exercise, was present in the field when this tragedy happened. He organized the dispatch of the medical teams and when all the casualties were taken care of and were ready to be flown to hospital he left with his helicopter, to report to the government.

A newspaper article accused Barak of fleeing from the scene of the tragedy. The words "Ehud Fled" were shouted from the rostrum of the parliament by one of the members of the Israeli cabinet.

For a number of years this slanderous accusation of Barak continued to appear periodically in various newspapers. I do not recall any of his friends and colleagues coming out loud and clear to remove this terrible accusation. Finally the State Comptroller was asked to make an investigation and review all the relevant facts. Only in early 1999 the report came out. The report stated categorically that there was no ground whatsoever for this accusation.

I do not compare myself in any way to Ehud Barak. But reading about his suffering from the newspaper slander reminded me of my own case.

Alan Challis wrote the following letter to Dr. Dinstein, following the newspaper attacks:

FIRST NATIONAL HOUSE
FINSBURY PAVEMENT
LONDON EC2P 2HJ
01-6382855
18th December, 1974

Dear Dr. Dinstein,

Zvi Alexander has mentioned to me that some questions are being asked in Israel as to the reasons for OMCO's Management continuing under the new ownership. Zvi is naturally concerned about this, and for my part I must confess that it seems a very strange question, to say the least.

As you know, when one buys a company, one of the predominant factors is management, and indeed this factor is often more important than the assets being acquired. In our case the problem was more serious because (1) as you know, our Bank did not have any oil expertise, and (2) a condition precedent by the Bank of England in approving the transaction had been that we acquired experienced management immediately to run the company.

We had a problem with Zvi in that up to the date of closing he had refused to commit himself as to whether he would agree to continue to manage OMCO. We were so worried by this fact that you may recall that our letter of commitment prior to closing was given in escrow to Mr. Ephrat on condition that it was to be released only if there was full agreement on the part of INOC and your Government for Zvi and his team to continue to serve the company. Only after Mr. Ephrat advised us of your agreement to these conditions, was the letter released to you.

While this correspondence with you was going on we were assuring the Bank of England that we would have professional management for the company, namely, the Signal directors and the existing management team. We would have been in a very awkward position with the Bank of England had he refused to continue his assignment - indeed, so far as we were concerned we would not have proceeded had Zvi refused to commit himself to us.

In view of Zvi's remark to me, I thought I should write you this letter

With best wishes,

Yours sincerely,
ALAN H. CHALLIS
Deputy Chairman.

* * *

When I returned to London on New Year's Day, Rachel, my wife, told me that Pat Mathews, the Chairman of FNFC had phoned several times with the question: "What happened to Burmah?" She obviously did not know and could not give him any answer.

Burmah Oil was primarily a holding company owning diverse assets, the largest and most valuable of which was the 23 percent ownership in British Petroleum. During the early 1970s Burmah Oil's management was talked into acquiring a fleet of tankers for oil transportation by a Greek ship owner named Kalakundis. They borrowed hundreds of millions of dollars for that purpose. When they purchased Signal Oil and Gas Company they borrowed an additional sum of four hundred and fifty million dollars from Chase Manhattan Bank to finance the purchase price. The security for that loan was their shares of British Petroleum. The covenant of the loan was that, as long as British Petroleum shares traded above a certain price, the loan was safe. If the price of the shares fell below that threshold, Chase could call in the loan.

Shortly before Christmas 1974 the stock market in England collapsed and the shares of BP went below the threshold specified in the loan agreement. Chase called in the loan.

Burmah Oil went to the Bank of England for help. The Bank of England, with the connivance of the British Government said: "Sell us your British Petroleum stock and we will pay off your loan from Chase." Burmah Oil had no choice but to agree.

It was actually a rape of Burmah Oil. This was the company that started the British involvement in oil. They founded the British Petroleum Company, which was one of the great contributors to the British success in World War I, by supplying fuel to the British Navy.

For many years after that "rape," law suits continued, instigated by angry Burmah shareholders. But the damage was done.

The mighty Burmah Oil, with which FNFC had been so proud to be in partnership, surrendered. Eventually Burmah Oil left the oil business altogether and remained in lubricants only.

On Lortcher's (Signal's) private plane. From right, Rachel Alexander; Aida (Lortcher's companion); Dutch Lortcher; and Zvi Alexander. Enroute from New Orleans to Los Angeles, 1972

At John Mecom's oil field in Louisiana. From left: Zvi Alexander; John Mecom, one of Signal's engineers; Bill Thompson, Executive V.P. of Signal, 1972

John Mecom and Zvi Alexander. Lake Charles Louisiana, 1972

Dutch Lortcher and Teddy Kollek, the Mayor of Jerusalem, during Lortcher's visit to Jerusalem's municipality, 1972

A Geisha party arranged by Sumitomo in honor of Signal's delegation. Seated from left: Bill Miller, Signal's V.P. Legal Affairs; Peter Rainier, V.P. Exploration; Geisha; Dutch Lortcher, the Chairman of Signal; Geisha; Zvi Alexander; and Signal's representative in Japan. Tokyo, 1972

Buz Ivanhoe, the geologist who initiated the Ziqlag project, with Zvi Alexander, approximately twenty-five years after the drilling of Ziqlag 1. London, 1990

The two "Thompsons" (not related), partners in creating A & T Exploration in 1989. Seated: Alf R. Thompson (then almost hundred years old). Standing: W.H. Thompson Jr., the former president of Signal and MAPCO.

From left: Dave Norman, an American consulting geologist; the Siberian Regional Governor; Miriam, the interpreter; Zvi Alexander; Boris Markelov, Regional Director of Development. Tomsk, Siberia, 1990

Dave Norman, the geologist; Zvi Alexander; and a representative of the
local government oil company. Baku Azerbeidzan, 1990

Visit to London by the Russian partners of the diamond
venture in Siberia, Mid-1990s

Zvi Alexander with A&T's local representative, Captain
Ye Myint Tun and his wife, Rangoon, Burma, 1996

Award ceremony of the "Large Gold Medal" to Zvi Alexander at the International
Stamp Exhibition, on the 50th Anniversary of the State of Israel, for his "Turkish
Palestine" postal history collection. From left: Zvi & Rachel Alexander, Hana &
Kobi Alexander, Tel Aviv, 1998

Signing ceremony of the oil concession in Burma,
Zvi Alexander. At his right, Dr. Zevi Dinstein, a partner in the
Burmese venture 1996

Our daughter, Dr. Shaula Alexander Yemini and her family

Our son Kobi Alexander, his wife Hana and their children, Jordan, Sharon & Ella

Chapter Twenty-nine

Warren Buffet

At the beginning of 1976, I came to the conclusion that there was no point in staying with the bank any longer. We should start thinking about leaving the sinking ship.

Adi Ephrat visited me in February of that year, and we discussed my plans for the future. He picked up the telephone in my office and called Dov Ben Dror in Tel Aviv. Ben Dror had been my chairman at I.N.O.C. and a very good friend. He was then the chairman of the Eilat-Ashkelon pipeline, which carried Iranian oil to the Mediterranean Sea. As mentioned earlier, 50 percent of this pipeline were secretly owned by the Iranian National Oil Company.

Adi filled Ben Dror in on my situation. Ben Dror was adamant that I should return to Israel as soon as possible. He said that the country needed me. When Adi returned to Israel, he went to see Dr. Dinstein to update him also. Dinstein made the same statement as Ben Dror. I should come back, the country needed me.

In March, the bank negotiated severance pay with all the Israelis. We all had five-year contracts. We were offered and accepted two years' salary as severance pay. In retrospect, I believe that we could and should have demanded some sort of pension arrangement as well. The termination of our employment was not our fault. We had severed our former ties in our country and had had to start afresh. Had I been alert enough to the pension problem, I think the bank would have responded positively. But I was not alert enough.

My wife and I left London at the end of March. Before going back to Israel, we first travelled to the States to say good-bye to all our friends. We went to New York, Houston, Denver, Las Vegas, San Francisco and back to New York.

For the first time in my life, I had almost two hundred thousand dollars in my possession, which included one hundred and thirty thousand dollars of severance pay and almost seventy thousand dollars saved from my salary during my service with the bank.

Wondering how to invest this windfall safely and profitably, I remembered a book I had read by a writer whose pen-name was Adam Smith. In his book he described the crazy stock market situation in New York in the late 1960s. As this was the period when I first tried to raise money in New York, the scene described by Adam Smith, whose name was actually Goodman, was very familiar to me. The only sane and prudent person mentioned in the book was a young man from Omaha, Nebraska, named Warren Buffet. The half chapter devoted to him told the story of a family fund that Buffet created in the mid-1960s and of his investment philosophy which was, in a nutshell, to buy undervalued assets at as cheap a price as possible. "Don't pay attention to the stock market gyrations. Remember that, being an investor, you are in partnership with 'Mister Market.' Your partner is sometimes in euphoria and at other times in depression. If you buy a good company, with good management, when 'Mr. Market' is depressed, you cannot go wrong." I was so impressed with this investment philosophy that, when my son Kobi was studying for his M.B.A. degree in New York, I gave him Goodman's book and wrote inside: "Kobi, Warren Buffet is my hero."

Goodman relates in his book that, at the end of 1968, Warren Buffet was so dismayed at the excesses of the stock market that he decided to terminate his fund. He distributed the accumulated profits to his investors, and he himself walked away with something like twenty million dollars, which was his part of the fund's success. He told his investors in his parting letter that he did not understand the market any more and he wanted to go back to Omaha to watch his children grow up.

I therefore thought that Buffet was no longer involved in the investment business. Still, I wanted to get hold of him and ask for his help in wisely investing my savings. I do not remember how I obtained his telephone numbers, but I still have them. I called his office, and a secretary replied that Mr. Buffet was away and would not be back for two weeks. As I was travelling around, I forgot to contact him again later. The fact that, if Warren Buffet had an office and a secretary he must be in business again, did not immediately register with me.

I found out much later that he was very much back in the investing business. He had created Berkshire Hathaway Inc., which is the most successful investment vehicle in the world. Today, Warren Buffet is the second richest man in the world, after Bill Gates. His fortune is in the neighborhood of forty billion dollars.

Had I pursued my hunch of trying to track him down, in 1976, when very few people knew of his existence, the two hundred thousand dollars of my savings which I would surely have entrusted to him would be worth today in excess of one hundred and fifty million dollars.

Several years ago, in 1996, I wrote a letter to Warren Buffet telling him the story which I have related above. I received a very nice letter back. My only consolation is that at least I had the right ideas.

I remember this "American farewell trip" as a very happy period in our lives. Finally we were free from the bank and away from the depressed mood prevailing in England. We looked forward with excitement to our return to Israel.

Ken Bialkin, our friend and the prominent New York lawyer, wanted me to meet two investment bankers. One was a senior partner in Goldman Sachs and the other banker represented a different investment house. I did not realize at the time that what Bialkin had in mind was for me to join one of these banking houses as a partner. It just did not dawn on me. Both meetings with the bankers were very pleasant. However, instead of saying that after being engaged in exploration for the last twenty years, I would like to use my skills in putting together oil deals, from the banking side, I emphasized the fact that I had an Israeli passport that prevented me from travelling to many oil-producing countries. Obviously this statement was counterproductive and I was not offered a partnership.

Another recommendation for a senior position in Israel was made by Dr. Yossi Vardi, who was then the Israeli commercial attaché in New York. He recommended to the American owners of the Israel Petrochemical Industries to engage my services as the president of their company in Israel. I met with their chairman of the Board in New York. We decided to meet again in Tel Aviv. At the end, he hired the former governor of the Bank of Israel for that position. Obviously the ex-governor had much more political pull than I, and that was of much greater importance in Israel.

Finally, we came back to Tel Aviv. I went to see Dr. Dinstein to inquire what position he had in mind for me. Nothing was clear. He mentioned this, that, and the other, but nothing concrete. He asked for my patience.

It was hot and humid. Everybody was working, but I had nothing to do. This situation continued for many weeks. In the meantime, my wife,

Rachel, agreed to become the principal of the Zionist Organization of America (ZOA) *Ulpan* (school for teaching Hebrew to adults, usually new immigrants). Our daughter and her husband were studying for their doctorate degrees at the University of California, Los Angeles, and our son was finishing his studies at The Hebrew University in Jerusalem. It was a strange feeling, being unemployed while everyone else was busy. I realized that I had made a mistake by not checking in advance the meaning, in practical terms, of the statement made by both Ben Dror and Dinstein: "Come back immediately, the country needs you." I should have gone to Israel for a short visit to find out what was being offered and what position was waiting for me, before we packed up, closed our apartment in London and sent all our belongings back to Israel. I told Dinstein on one occasion that he had caused me the unnecessary expense of tens of thousands of dollars created by our hasty return to Israel.

Chapter Thirty

The Life of an Independent Oilman

My career as an independent oilman, which continues to this date, more than twenty-five years later, started in the latter part of 1976 without my having planned to embark on such a career.

To describe all my various enterprises in so many countries would fill another book, which I am not yet ready to write. Instead, I will only mention some of the more important ventures in which I was involved during this long period.

Whilst we were still in Israel in the long summer of 1976, Adi Ephrat mentioned to me that a lawyer friend of his had a client who was heavily involved in Guatemala. Adi knew about my interest in that country and recommended that I should meet the attorney and his client. He was none other than Professor Yaacov Neeman, the most prominent tax lawyer in Israel, and a future Minister of Finance. His client was an ex-Israeli who lived in Mexico, by the name of Mordechai Katz.

I went to meet Katz in Zurich. We had a long chat, at the end of which he said that oil was really not his field, he did not understand it and therefore was not very much interested. By coincidence, when I met his wife, I realized that she had served in my army unit, back in the early 1950s.

When I returned to Israel, I found an urgent message to go and see Professor Neeman. When I met him, he handed me a plane ticket to Guatemala City and said that Mordechai Katz would be waiting for me there on the following Sunday.

I arrived in Guatemala City on that Sunday, and the following afternoon, I was sitting in the presidential palace with the President of Guatemala. I drew maps for him to explain why I thought that the Mexican oil trend, where large oil deposits had been discovered recently,

251

continued into his country. The president was a very intelligent and erudite man. He was of Norwegian descent and had a Norwegian-sounding name. He was a graduate of West Point who spoke perfect English.

Bill Thompson was going to move in September 1976 from the presidency of Signal Oil, which became Burmah Oil USA, to be president of MAPCO (Mid-America Pipeline Company). MAPCO was a very successful energy company with headquarters in Tulsa, Oklahoma. It owned the largest pipeline of LPG (liquid petroleum gas) in the United States. MAPCO was a relatively young company with a very interesting history. In America, before a pipeline is laid, the pipeline company needs to buy the right of way from the hundreds of farmers and other landholders through whose land the pipeline will pass. The founder of MAPCO, Bob Thomas, found a brilliant solution to this obstacle. He reached an agreement with the Katy Railroad Company whose tracks ran from Texas to Minnesota. MAPCO laid its pipeline along these tracks, without having to negotiate and purchase the right of way from thousands of land holders. This was almost a license to print money. By the time Bill Thompson became the president of MAPCO, the company had, in addition to the pipeline, large coal mines that are to this date the most successful and profitable ones in the United States. MAPCO also owned two oil refineries and had a successful oil exploration and production department.

Bill Thompson shared my enthusiasm regarding the prospects of oil exploration in Guatemala. We therefore had a willing, ready and able buyer for the Guatemala project.

Although Bill Thompson was the president of the company and was very interested in the Guatemalan opportunity, we still had to negotiate the deal with the correct person in the organization, in order to prevent a mishap similar to the one I had with the head of Patino Mines, back in 1969, that I have described in Chapter Six. The deal had to be negotiated with, and agreed to, by Lee Wright, MAPCO's vice-president of exploration. As Lee knew that I was a friend of Thompson's, I had to neutralise his resentment and opposition to Thompson's involvement. My friend, Barth Walker, the oil lawyer from Oklahoma City, who was previously the chairman of Mayflower, travelled many times to Tulsa to discuss the Guatemala deal with Lee, without my participation in any of those meetings. When, finally, Lee was pacified, I appeared on the stage and we signed the agreement with MAPCO. According to the agreement, Mordechai Katz and I were going to be paid a large sum of money by

MAPCO upon the signature of the contract with the Guatemalan Government, plus an overriding royalty from any oil which MAPCO would discover and produce there. An overriding royalty entitles the holder to receive a certain percentage of the oil produced from the licensed area, as long as the license is in force, without having contributed any investment.

I spent many months travelling back and forth between Oklahoma and Guatemala City. Bob Thomas, the chairman and creator of MAPCO, and Bill Thompson, the president, visited Guatemala City several times. They met with the president and the minister of mines and everything looked wonderful. As there was a rule in MAPCO that the chairman and the president should not fly in the same plane, each such visit included MAPCO's two Falcon jets, which we jokingly called MAPCO's air force.

In the spring of 1978, Shenandoah Oil, a small American oil company which had a producing oil field in Guatemala with proven reserves of fifty million barrels of oil, suddenly made an official announcement to the American Stock Exchange that its proven reserves in Guatemala should be revised and downgraded by 50 percent. It was an unexpected, severe blow to the Guatemalan project, in view of which Bill Thompson had to reconsider MAPCO's involvement. He hired a well-known geologist by the name of Phil Chenovitch to reappraise the Guatemala project. After a month Chenovitch submitted his report. In essence, it stated that, in his view, there was no geological connection between the prolific Mexican oil trend and the Guatemala area. MAPCO had no choice but to abandon this project. Phil Chenovitch has been right so far. In spite of numerous attempts by various companies in Guatemala during the last thirty years, no substantial oil reserves have been found there.

* * *

Another venture in which I started to be engaged at the end of 1976, and which continued through 1977 and 1978, was buying oil-producing properties in the United States. I was involved with one large German group pursuing such projects in many parts of the United States. There was a very generous tax regime in Germany which favored such foreign oil exploration activities.

In view of my involvement in these two projects, Guatemala and the German-US production and exploration activities, I spent most of my time in the US and Guatemala. I told my wife in mid-1977 that it was too difficult and too expensive to come for weekends from America to Israel, and that we should consider moving back to London at the end of her school year. Our son was then going to New York to study for his M.B.A. degree. It meant that both our children would be living on the other side of the world. I therefore suggested to my wife that we should move back to London, which was a hub of the oil industry and located halfway between Israel and America. We moved back to London at the end of October 1977, and we are still there over twenty-five years later.

To this day, we have kept our house in Zahala. We also bought a beautiful apartment in Tel Aviv, where we spend several months a year. I searched for a long time for a suitable apartment with twenty-four hour security and doormen. We finally found such an apartment on the thirteenth floor in Migdalei Pinkas in Tel Aviv, where we can enjoy the view of the Mediterranean Sea from every window.

* * *

In late November 1977, during one of my visits to New York, I went to see Dr. Zevi Dinstein who had by then been nominated to be Israel's economic minister in the USA. Dinstein asked me whether I had already met Yitzhak Modai, the new Israeli minister of energy. There was a new government, headed by Menachem Begin, which for the first time had instituted a Ministry for Energy. When I replied that I had never met Modai, he advised me strongly to cancel my proposed flight back to London that evening, and to meet Modai who was coming to lunch with Dinstein the next day. We had a long lunch together, and when Dinstein had to leave at 3 p.m., Modai asked me to stay on. We parted at the end of a four hour meeting, after 5 p.m. I was extremely impressed with Yitzhak Modai, a graduate engineer, and an enormously intelligent and bright man. It was a pleasure to be with him.

During our long conversation, Modai asked my advice about the five or six drilling rigs of Lapidoth, which were standing idle, accumulating rust and costing maintenance and insurance to boot. I replied that the easiest solution would be to advertise in the *Oil and Gas Journal*, asking for sealed bids from interested buyers. But a much more creative and profitable solution would be to find a drilling contractor in the United States, where at that time there was a shortage of drilling equipment and embark on a joint venture with such a contractor. Thus we could put the rigs to work, train some of the Israeli crews in America without any cost to us, and earn an income from the rigs' drilling operation. Modai liked the idea of the joint venture very much and asked me whether I would be willing to help him. Obviously, I answered in the affirmative.

During the second week of December, in London, I was debating with myself whether to go to the States. The Christmas holiday season was approaching and everybody was getting ready to go on vacation. I phoned Modai in Jerusalem and asked him how serious and urgent his request for assistance in the Lapidoth rig problem was, as I was wondering whether to postpone my US trip to the beginning of January. Modai asked if I could possibly go right away, which he would greatly appreciate. I flew to Oklahoma City and asked Barth Walker to recommend a contractor for the drilling venture. Barth phoned one contractor who said he was not interested while the second one, Raymond Hefner, the owner of Bonray Drilling Company replied that he wanted to meet me and discuss the subject. Bonray was an amalgamation of his and his wife's names, Bonny and Raymond. We met and Raymond Hefner agreed to come to Tel Aviv during the first week of January 1978 to see the equipment and review the proposed project.

When, as promised, Hefner came with his assistants in the first week of January, we all met in Modai's office and the principles of the joint venture were agreed upon.

Another attack on me personally in the Israeli papers was instigated, this time by Brigadier General Israel Lior. Lior had replaced me as the head of the Israel National Oil Company, after we left for England in 1974. He was the former military secretary of the late Prime Minister of Israel, Golda Meir. Lior resented the fact that Modai asked me for my advice and help in solving the rig situation, which he had grossly mismanaged. By that time Lapidoth was a subsidiary of I.N.O.C., for which Lior was responsible. Probably in order to divert the wrath of Modai over his incompetence, he decided to call in a journalist, one of the most vicious in Israel, to publish a slanderous article about my continuous bad influence on the Israeli oil industry, this time relating to the Lapidoth rigs. Other newspapers picked up the story with another muckraking exposé. One of the most slanderous articles was published during the Passover vacation while we were visiting Israel. I was so offended and incensed that, although we had tickets to go back to London on El Al two days later, at the end of the holiday, I decided to buy new tickets and leave the same evening on an SAS flight.

* * *

The contract with Bonray was signed in the summer of 1978. The idle rigs of Lapidoth were shipped to the US and started working there immediately. In addition, Israeli crews received American training on those rigs for the joint venture. It was a very profitable contract for Lapidoth and Israel, and everybody was very happy—except General Israel Lior. Mordechai Chen, the president of Lapidoth and my former employer, came to Oklahoma City to visit his people and his rigs. We were the best of friends during that visit. Ever since I had left Lapidoth, twelve years earlier, we had kept in touch and I had also helped him in other endeavors.

* * *

After the completion of the drilling rigs transaction, I brought before Raymond Hefner several other projects that we considered to explore jointly. A native of the southwest, Hefner originally trained to be a certified public accountant. He entered the oil business and by hard work and perseverance built a successful oil company—Bonray. The many millions of dollars he made did not change him. He remained the same wise and considerate human being. I learned to know, love and respect the man. He was the salt of the earth, the "beautiful American." He was wise, quiet, thoughtful, considerate, and a very good friend. He later merged Bonray into a publicly traded company called ANTA, where he became a controlling shareholder with a 40 percent interest. At one stage the idea was mooted that I join ANTA permanently to establish and develop their international division. The plan did not materialize, as the idea was too revolutionary for ANTA, whose entire experience and involvement was in US domestic activities.

* * *

Another matter on which I dealt with Modai was the supply of coal for the generation of power in Israel. At that time a large power station was being built in Hadera—the first to use coal instead of fuel oil, as used in all other power stations. As I have mentioned before, MAPCO was the owner of the most successful and profitable coal mines in America. Here, again, my idea was to form a joint venture with MAPCO and not only to buy coal but to produce it and be part owner of the mine. The subject of cooperation with MAPCO was presented to Yoske Tulipman, the president of Israel Electric. He went to see MAPCO's coal mines and was enthused by what he saw. A long term supply contract was agreed upon, and there remained only two points to be straightened out. The first was Israel Electric's insistence that the contract should be signed by MAPCO, while MAPCO wanted to have one of its subsidiaries as signatory. The reason was that MAPCO's refineries were buying oil in Libya, and they were afraid of political repercussions if MAPCO's name would appear as a supplier of coal to Israel. The other point was that Israel Electric wanted a five-year, fixed price contract, while MAPCO preferred to use the prevailing market price. (How right MAPCO was. The price of coal went down by almost 50 percent the following year.) Tulipman insisted on conducting the negotiations with MAPCO by himself, without my participation. He did not advise me of these two points of disagreement, which I am sure I could have solved during one short visit with Bill Thompson. As a result, the contract with MAPCO never came into force.

* * *

Another very important subject that I discussed with Modai was the possibility of supplying North Sea oil to Israel. A unique opportunity arose in February 1979 to buy a part of the Ninian oil field, one of the largest oil fields discovered in the North Sea. It had proven producible reserves of almost two billion barrels of oil. For political reasons, such a transaction obviously had to be fronted by a third party—the Rothschilds for example. That would not have been too difficult to arrange.

Michael Belmont, a prominent London banker and a partner in Cazenove, one of the oldest and most prestigious banking houses in London, in existence for hundreds of years, was a close personal friend of Bill Hutchinson, the banker who had helped me, back in 1973, to raise money in England. I first met Belmont in 1973, and we continued meeting periodically.

Michael Belmont was the creator of London & Scottish Marine Oil Company Ltd., which later became LASMO, one of the largest and most successful independent British oil companies. LASMO had an 8 percent participation in a group which discovered the Ninian oil field in 1974.

Our story begins in early 1979 when I met Belmont to discuss the possibility of purchasing part of LASMO's interest in the Ninian oil field.

Their interest was held in two entities, one holding 6.4 percent and the other 1.6 percent. He therefore could dispose of the 1.6 percent without many legal difficulties. They urgently needed more than one hundred million pounds to finance their share of the development costs of the Ninian oil field. It was therefore an opportune time to discuss such a purchase. This unique chance existed for one month only, during February 1979. At that time, LASMO did not yet have an office, nor any employees, and Michael Belmont kept everything in his head. He was LASMO. I called Minister Modai in Jerusalem while he was participating in a parliamentary meeting. I considered this possibility so momentous that I asked for him to be called out of the meeting, and he came to the phone. Unfortunately neither Modai nor the Israeli Cabinet rose to the occasion.

This was the period following the fall of the Shah of Iran, when oil prices were rising sky high. Had this golden opportunity been seized upon, it would have saved Israel hundreds of million of dollars and a lot of grief.

* * *

One of my personal experiences, which will illustrate the crazy market conditions and the soaring prices of oil which followed the fall of the Shah of Iran, is worth mentioning. One afternoon, in February 1979, I received a telephone call from Marvin Billet, the New York lawyer and our partner in Ghana. Marvin told me that our mutual friend, Cletus Wotorson, the former minister of mines in Liberia and the present managing director of Liberia's oil refinery, was running out of crude oil. Could I help? I phoned Mark Rich's office in London and was told that they had two oil tankers presently sailing around the Cape of Good Hope. One held 125,000 tons of oil and the other 100,000 tons. (There are seven barrels of oil in one ton).

Each ship could be diverted to Liberia and the price would be $16.75 per barrel. I phoned Billet with this information but, in view of the time difference with West Africa, he asked for an option till the following morning. I spoke to Mark Rich's office again and was told that the situation was so volatile that they could not give us an option. First come, first served. The next morning I received confirmation from Liberia to buy one of the tankers. When I called Rich's office, I was told that they had sold the 100,000 ton cargo during the previous night and had decided not to sell the second cargo and to let the second tanker continue sailing to Houston.

Six days later, Bill Thompson, the president of MAPCO, whose company's two refineries were also in great need of oil supplies, told me in a telephone conversation that a 125,000 ton tanker was then unloading in Houston and the price of the oil was $24.75 a barrel. As it takes six days of sailing from the Cape to the Gulf Coast of America, I figured out that it must have been the same tanker that we were discussing with Mark Rich's company in London. Their decision not to sell it six days earlier had brought them an additional profit of almost seven million dollars. ($8.00 additional price per barrel, multiplied by seven barrels per ton, multiplied by 125,000 tons.) One must also remember that the price of $16.75, which they had quoted us in London, must have included a nice profit already.

Those two mad years of 1979–1980 made Mark Rich and his partner Pinky Green billionaires. Their names appear in a list of the four hundred wealthiest Americans published in the year 2000.

The story of Cletus Wotorson, who had needed the oil for his refinery in Liberia, is also an interesting one. Cletus was one of the first Africans to graduate in geology at an American university. He was appointed by President Tubman to be the minister of mines in Liberia. In the mid-seventies, there was a military coup in Liberia; President Tubman was deposed and all the ministers were taken to the beach to be executed. Sergeant Doe, who had organized the coup, came to watch the "ceremony." When he saw Cletus tied to a wooden pole, he ordered his release. During President Tubman's term of office, Sergeant Samuel Doe had been sentenced to hard labor, which he served in one of the mines. Minister Cletus Wotorson visited this mine one day, and, being in need of a storeman who could read and write, arranged the release of Samuel Doe from his hard labor and put him in charge of the tool warehouse. This appointment saved Cletus' life after the military coup.

∗ ∗ ∗

To continue the story of LASMO and the missed opportunity of buying an interest in the Ninian oil field: A month after our initial discussion LASMO already had its first employee, a managing director called Hector Watts, a former Shell Company executive. Watts was no longer willing to consider the sale of part of the Ninian field, but he was ready to discuss a firm commitment to sell approximately twenty million barrels of oil over the following five years, 1980–1984. Again, the opportunity was missed, although I had arranged a commitment of a hundred million dollars for the Israeli government to finance such a purchase. The four letters relating to these lost opportunities are shown in the following pages.

CAZENOVE & CO.

L. MEINERTZHAGEN

J. R. HENDERSON
G. J. CHANDLER
M. J. K. BELMONT
M. J. de R. RICHARDSON
A. J. S. COOMBE-TENNANT
E. L. WINDSOR
G. D. WENTWORTH-STANLEY
J. KEMP-WELCH
R. L. H. LYSTER
G. S. P. CARDEN
A. D. A. W. FORBES
I. A. D. PILKINGTON
LORD FARINGDON
P. J. SMITH
D. L. MATHEW

P. B. MITFORD-SLADE
C. D. PALMER-TOMKINSON
U. D. BARNETT
H. de L. CAZENOVE
D. J. ROCHESTER
A. F. BAMFORD
T. SCHOCH
H. A. LOVEDAY
P. J. SCOTT-PLUMMER
R. B. SMITH
D. C. GODWIN
C. J. CAZALET
A. H. J. MUIR
N. A. GOLD
B. E. A. PASCOE

TELEPHONE: 01-588 2828

*12 Tokenhouse Yard
London EC2R 7AN*

MJKB/JK

9th February 1979

Dear Mr. Alexander,

   It was a pleasure to see you again today. I thought I would just confirm that in the brief and unofficial discussions we had, that the client for whom you are acting would like it to be known to LASMO that they are potentially interested in a purchase of any part of their interests in the Ninian Field, and I have reflected this to the company, but that we await the necessary figures on which to evaluate any potential deal, and I will contact you as and when these are available.

   I also wish to emphasise that other companies have already approached LASMO with similar interests and that the company has told them, as indeed I am telling you, that until figures can be evaluated nobody has any preemptive position.

   I look forward to further meeting you on your return from the United States.

      Yours sincerely,

Z.Alexander, Esq.,
4 Raynham,
Norfolk Crescent,
London, W.2.

TELEGRAMS AND CABLES: CAZENOVE, LONDON, E.C.2.   TELEX: 886758

GREYHOUND INTERNATIONAL FINANCIAL SERVICES LIMITED

Managing Director's Office

5 Grafton Street London W1X 3LB Tel:01-629 1208 Telex : 22465 Cables : Greyint London W1

Zvi Alexander Esq
4 Raynham
Norfolk Crescent
London W2

12th February 1979

Dear Zvi,

Following our various conversations I am pleased to inform you that we, together with a small number of associated banks, are willing to arrange a medium term loan of up to U.S. $100 million to enable the Israeli government or any of its agencies to purchase oil production in the North Sea. The loan which could have a life of up to 8 years should inter alia, be guaranteed by the government.

It is obvious that at this time it is difficult to present you with the exact cost but in principle I can say that the rate of interest on the loan will be based on 3 or 6 months LIBOR and that it will be priced at keen market rate.

I hope that your negotiations in Israel will be successful and I am looking forward to doing business with you.

Yours sincerely,

David Nussbaum

DN/eek

Directors G H Trautman Chairman (USA) D Nussbaum Managing C Franssens (Belgium) S G Marias (USA) R S Grim (USA) V. Hollard (France)
Registered in England at above address Registered number 1154048

London & Scottish Marine Oil Company Limited

Bastion House,
140 London Wall,
London EC2Y 5DN.
Telephone 01-600 8021
Telex 8812970

HW/SDL   <u>Confidential</u>   6th March 1979.

Mr. Z. Alexander,
4 Raynham,
Norfolk Crescent,
London W.2.

Dear Mr Alexander

I was very pleased to meet you yesterday with Michael Belmont and as promised am now writing to you to summarise the main points of our discussion.

As you know LASMO presently holds just under 8% of the Ninian Field which is one of the largest in the North Sea. The Field came into production in December and we are only now beginning to reap the reward of the many years of effort which LASMO has made to finance their share of the Ninian Field.

As you can well imagine therefore, the idea of disposing at this moment of part of our interest is not one which commands universal agreement in our Board.

Obviously the value put on a share of oil in the ground in a proven field in the North Sea will be a material factor in any ultimate decision.

The valuation of the oil in the ground for an acquirer is a relatively simple calculation given certain agreed assumptions as to reserves, production profile and costs, but these have already been largely established and agreed by the Ninian Partners.

Much more complicated is the assessment of the impact which the disposal of part of our interest would have on our Company and its operations, bearing in mind that the burden of the loan finance and oil production stock would then remain to be borne by the Company's retained interest.

In addition we have to take into account the view expressed by the operators that the total recoverable reserves are likely to be substantially in excess of the declared 1.2 billion barrels, whilst the better than expected performance from the wells drilled so far equally tend to the view that the peak production rate will exceed the rate presently foreseen in the agreed profile.

continued....

- 2 -

<u>Mr. Z. Alexander.</u>   6th March 1979.

I have, for our own purpose, put in hand a through going
exercise in order to establish these various values with
a variance to assess sensitivity on certain key factors
and I hope to have these completed in time for our next
Board Meeting.

In addition to the possibility referred to above, you also
raised the idea of a long term crude oil supply agreement.
As I advised you we would be prepared to consider this
provided, however, it were accompanied by a prepayment this
year, which could be made in a number of instalments, to
compensate us for the loss of freedom which would result
from committing our crude oil availability over a number
of years.

We can make available over the period 1980/84 a total of
2 million tonnes which could increase to 2.5 million tonnes
over 1980/85. The crude oil would be sold at market prices
to be agreed quarterly, the prepayment we have in mind being
of the order of 25% of the value of the contract.

As you pointed out, these two possibilities are not
incompatible and a crude oil deal could well be incorporated
at some stage into a longer term arrangement such as the one
you first mentioned.

I hope this summarises our discussion and that it will
enable you to give your Principals a sufficiently clear
picture of the situation.

I will get in touch with you as soon as we have considered
this question with the Board but in the meantime it would,
of course, be helpful to know of any further reactions from
your Principals to the ideas mentioned above.

Our discussion and this letter is naturally without commitment on either side at this stage and I can confirm that we
are not engaged in any similar discussions with another
party.

With kind regards

Yours sincerely

(Hector Watts)

The Israel National Oil Company Ltd.
March, 1979

Mr. Zvi Alexander
London

Re: The LASMO transaction.

Dear Sir,

I received your letter dated 7 March 1979 only on the 20th of March, and
therefore my late reply.

Following the reply to your letter written by the Head of Oil
Administration , I understand that the future negotiations for of the
pre-purchase of part of LASMO's oil production will be handled by him,
and I hope that he has approached you already about this matter.

The Minister of Energy decided to entrust the negotiations for the
purchase of a participation in the Ninian oilfield to Israel National Oil
Company. I would be therefore grateful if you would keep me informed of
developments in this matter (after the Board meeting that is to take place
on March 20th 1979).

> Sincerely
> Elazar Barak
> I.N.O.C.
> Managing Director
> 25 March, 1979

Copy to: Mr. Itzhak Modai – Minister of Energy and Infrastructure
          Mr. Shimon Gilboa – Director of Fuel Authority

חברת הנפט הלאומית לישראל בע"מ

תל-אביב · ת. ד. 20115 · בית מעיא, דרך פתח-תקוה 74 · טל. 36113, 37934

HE ISRAEL NATIONAL OIL CO. LTD

-AVIV, P.O.B. 20115 · MAYA BLDG., 74 PETAH TIKVA RD. · TEL. 36113, 37934

25 במרץ 1979

לכבוד
מר צבי אלכסנדר
לונדון

א.נ.,

הנדון: עסקה   LASMO

את מכתבך מיום 7 במרץ 1979 קבלתי רק ביום 20 במרץ 1979, ומכאן האיחור
במתן תשובתי.

על פי ההיחסותו של מנהל מינהל הדלק למכתבך האחרון הבנתי כי המשך הטיפול
בנושא קניה מראש של תפוקתה החפשית של LASMO יטופל על ידו, והנני
מקוה כי פניות אליך, על ידו, בתחום זה כבר נעשו.

שר האנרגיה והתשתית החליט להעביר המשך הטיפול ברכישת הזכויות בשדה
NINIAN לחברת הנפט הלאומית בע"מ. אודה לך על כן אם תעדכן אותי
לגבי ההתפתחויות בתחום זה (לאחר ישיבת מועצת המנהלים, שהיתה אמורה
להערך ב-20 במרץ 1979).

בכבוד רב,

דר' אלעזר ברק
מנהל כללי

העתק: מר יצחק מודעי – שר האנרגיה והתשתית
מר שמעון גלבוע – מנהל מינהל הדלק

אב/צב

CABLES: ISNEFT TEL-AVIV   מברקים: ישנפט תל-אביב   TELEX: 33498   טלקס: 33498

In April or May 1979, when it became clear to me that Israel would not rise to the occasion, I told Mark Rich's executives in London of this opportunity to buy the twenty million barrels from LASMO. They were extremely interested and hastened with me to several meetings with LASMO's management to try and negotiate a similar contract. By that time LASMO already had offices and staff, including an oil marketing executive. The price of oil was going up every day; as a result LASMO refused to consider any long term contract for the pre-sale of oil.

Had I been more commercially inclined, instead of worrying only about Israel's oil supplies and talking to Modai, I would have immediately approached both Mark Rich and MAPCO. They were both very much interested in securing long term oil supplies at official prices. Either Rich, or MAPCO, or both, would have concluded the transaction with LASMO within a week or two, and I personally would have made enough money to start a new oil company. But owing to my naivety, or perhaps foolishness, I was waiting for the Israeli Government's cumbersome wheels to start turning. Thus, a once in a lifetime opportunity came to nought.

The only explanation I have for my strange and uncommercial behavior is that I probably still had a strong feeling of being Israel's emissary, and that my duty was to look after the State's needs. From the time we moved to England in 1974 and until the early 1980s, I continued to be concerned with the mission of supplying oil to Israel—without being asked to do so, and obviously without any remuneration. I failed to realize that I was now a private individual and that it was my duty to devote myself to taking care of my family and myself.

Probably in my subconscious I still felt that I was Israel's public servant. Only in the early eighties did I became aware of the truth of the *Sayings of the Fathers*: "If I am not for myself, who will be for me?" The only person who would take care of my family was me. I believe that part of my new attitude was provoked by renewed attacks in the Israeli press instigated by my "colleagues" in the oil establishment in Israel. They were hoping that, with my move to England, they had finally gotten rid of me. Yet, here I was again with new ideas and initiatives disturbing their pleasant slumber.

* * *

The idea of bringing gas from Egypt to Israel was first mentioned to me by Yossi Honig, Minister Modai's economic assistant. During one of our meetings he said: "Zvi, with all your world-wide contacts, why don't you think about organizing a group to bring the gas which is wasted and flared (burnt) in Egypt to Israel to be used in power generation." This was a brilliant idea.

I spent quite some time and effort in creating such a group. Bill Thompson, the Head of MAPCO, ordered a feasibility study which showed the economic merits of such a scheme, and MAPCO agreed to head the project and to lay and operate such a pipeline. Frank Zarb, who was formerly the energy czar in President Nixon's administration, was then a senior partner at Lazard Frères, one of the oldest and most prestigious banking houses in the world. He agreed to arrange the financing for the pipeline.

In January or February 1980 a meeting was held at the Waldorf Astoria Hotel in New York. Minister Modai, Bill Thompson, Frank Zarb and I participated. What was needed was Egypt's agreement to supply the gas, which Frank Zarb was going to pursue with his good friend, the Egyptian ambassador in Washington, Mr. Ghorbal. The preliminary indications from Ghorbal were positive. In the end, the Egyptian government did not rise to the occasion. Egypt could have earned billions of dollars over the last twenty years, selling gas to Israel, which otherwise was burnt and went to waste. Then, in the year 2003, this opportunity was being very seriously discussed again. Egypt was now ready to supply any quantity of gas that Israel may desire to purchase.

Chapter Thirty-one

Discoveries in Egypt and Jack Bitton, the Egyptian Master Spy

In late summer of 1978, I was approached by two unconnected businessmen regarding concessions for oil exploration in Egypt. The first, Jack Bitton, introduced himself as an Egyptian Jew, who had left Egypt for Israel after the Sinai campaign in 1956. He had opened a travel agency on Brenner Street in Tel Aviv. Ten years later, Bitton went to Germany, where he married a German woman, and they had a young son. In Egypt, prior to 1956, he had worked for the Shell oil company. He had Egyptian friends who previously worked with him in Shell. They were now working in EGPC, the Egyptian National Oil Company, and helped him to obtain an oil exploration concession in the Western Desert of Egypt, in an area called Mleiha.

In the 1930s, Phillips Petroleum Company had discovered oil at Mleiha. The quantity was small, the prices were very low then, and this oil was never produced. Bitton succeeded in raising some private money in Germany to develop the Mleiha concession. By 1978 most of the money was gone, and he came to me for advice and help. He was told by a mutual acquaintance in America that I was the man who could help.

My advice was that he should make a deal with an oil company. Money by itself would not solve his problem. He needed oil exploration expertise as much as, or more than, money. He asked me to come to Germany and explain this to his investors. I went to Frankfurt, and had a meeting with the Board of Directors of Agipetco, Bitton's company. Bitton later informed me that his Board had decided to raise the necessary funds in Germany; therefore, my help was not needed. That is where our discussions ended.

In the same month as my first meeting with Bitton, an old friend of mine, Kivi Kounine, came to see me. He told me that he had met a Greek businessman in Athens who could obtain an oil concession in Egypt. "What does one do with an oil concession?" Kivi asked. "First get the concession and then I will tell you what to do with it," I replied. I explained to him various methods of compensation in a "farm out" deal (selling a specific percentage of an oil concession) such as cash payment, a carried interest, net profit interest, royalty, etc. It all depended on the attractiveness of the concession and on finding the right partner. We then agreed that Kivi would engage the services of John Christensen, my geologist friend from Rome, and I introduced Kivi to him. Christensen would choose a promising area and I undertook to find the oil industry partner.

As I was then working on several international oil projects with Raymond Hefner, who made the joint venture to use the drilling rigs of Lapidoth, I told Raymond about these two opportunities in Egypt. Raymond was interested.

John Christensen went to Egypt and recommended an area called Khalda which, by a strange coincidence, bordered on the east Bitton's Mleiha concession. The Khalda area had also been explored by Phillips in the 1930s and had been abandoned. John Christensen believed that it had serious potential. Medoil, Kivi Kounine's company, which was created for this venture, signed a letter of intent with the Egyptian National Oil Company for the Khalda area.

In October 1978, several months after our previous meeting, I received a call from Jack Bitton that his Board of Directors had come to the conclusion that they did need my help. Would I please find them an oil industry partner?

I suddenly found myself dealing with two adjoining concessions in Egypt, belonging to two completely unrelated parties. As I did not see any conflict of interest in dealing with both areas, I decided to pursue both opportunities.

Raymond Hefner came to London and met Jack Bitton who had come from Frankfurt, and they agreed on the principles of a deal between Raymond's company, Bonray, and Bitton's company, Agipetco. Hefner then met with Kivi Kounine and invited him to present his deal when Medoil would have an agreement with the Egyptian authorities.

I told Bitton that, subject to the conclusion of a deal with Hefner, my fee would be one hundred thousand dollars and a 2.5 percent overriding royalty of any oil produced from the Mleiha concession. Bitton agreed.

In November, Jack Bitton, his English consulting geologist Ronald McLean, and I went to Oklahoma City to negotiate a final agreement with Hefner.

During the three-day negotiating sessions with Hefner, Bitton, using the Middle Eastern methods of conducting business, raised his demands at each and every meeting. I was furious, while Hefner remained very cool. He listened attentively to Bitton's demands. When Hefner saw how upset I was, he patted me on my knee, and whispered, "Don't worry."

On the last day, before Bitton and his geologist were leaving for the airport, Hefner said: "Now, Jack, you can have the terms we agreed upon in London. No more and no less." Bitton almost fainted. All the goodies that he believed to have gained during the last three days suddenly vanished. Bitton agreed. Obviously, he had no other choice.

After Bitton and his geologist left, Hefner and I walked around in his office building. We passed the office of Raymond's brother, Kenneth Hefner, who was working for Bonray as a petroleum engineer. Raymond told his brother that we had concluded an agreement with Bitton. His brother replied: "Raymond, you are making a very serious mistake. Both I and the other petroleum engineer believe that there is no oil in Mleiha to justify our involvement." But Kenneth Hefner was disastrously wrong. Mleiha subsequently proved to be the second richest area in the Western Desert of Egypt. Hefner's Bonray could have become one of the wealthiest companies in the State of Oklahoma.

Kenneth Hefner's disparaging remark was a very serious statement. Bonray was a part of ANTA, a public company listed on the stock exchange. Although Raymond controlled 40 percent of ANTA, he still could not make unilateral decisions, especially against the advice of his professional people. The Bitton deal went on hold.

In the meantime Kivi Kounine's company, Medoil, had concluded its negotiations in Egypt and required a serious oil company to sign the agreement. They urgently needed a bona fide oil company for two reasons: To provide five hundred thousand dollars, the sum of the signature bonus which had to be paid to the Egyptian government, and to present a substantial oil exploration company with which Egyptian authorities would agree to sign an exploration concession.

The Medoil deal was relatively inexpensive compared to Bitton's Agipetco deal, as Khalda had no oil, whereas oil was already proven to exist on Bitton's Mleiha concession. It was therefore much easier for

Bonray to conclude a deal for the Khalda concession with Medoil. Medoil wanted half a million dollars for their efforts. I went several times to Oklahoma City to negotiate with Hefner. Hefner agreed to pay Medoil the half a million dollars, pay the Egyptians another half a million dollars as Signature Bonus, and to sign the concession in Egypt.

I went back to New York after concluding the agreement with Hefner. I stayed the night with my son who was then studying for his M.B.A. degree, at New York University. First thing the following morning, I called Kivi in London and told him about the conclusion of an agreement with Raymond Hefner. Kivi congratulated me profusely and then asked: "What about a royalty?"

I said: "You S.O.B., you never asked for a royalty. For three months I have been talking to Hefner on your behalf and the only thing that you asked for was the five hundred thousand dollars."

Kivi admitted that this was the case. He then added that when I first explained to him, a year earlier, the various benefits that may arise from selling a concession, I had mentioned a royalty as one of the alternatives. "Could you please try. You are such a good friend of Raymond Hefner's," he pleaded.

I called Hefner in Oklahoma City and repeated Kivi's request. "But you never asked for it," Raymond said.

"This is true," I replied, "they forgot to ask for it."

"What do you have in mind?" asked Raymond.

"Raymond, if there is no oil, 3 percent of a zero is still a zero. If there is oil, a 3 percent royalty would not make such a big difference either."

There was a long silence on the line. After the pause, Raymond said, "You have it." I picked up the phone, called Kivi in London, and told him about my achievement. Kivi was deliriously happy.

But my hasty call to Kivi in London cost me many millions of dollars. What I should have done, as a prudent businessman, was to sit down and think quietly about what would be an equitable division of this royalty between Kivi and me. After all, but for my friendship and special relations with Hefner, there would not have been any royalty. There was no reason in the world for me to have phoned Kivi immediately with the results of my achievement. Had I waited and considered the pros and cons of the situation, I believe that I would have come to the conclusion that I was entitled to one half, or at least one third, of the 3 percent royalty.

I failed to pursue the matter as I was very busy in South America and was spending most of my time in Peru. When, several months later, I wanted to summarize my interest with Kivi, he suddenly became very busy. Whereas before he had visited me almost every day and pestered me on the phone as well, it now became difficult to catch him. When we finally agreed on a meeting, one of his partners, whom I had never met before, came along and started negotiating with me. Kivi did not say a word throughout this three-hour meeting. It was a big mistake on my part to talk to his partner. I should have insisted that either the two of us, Kivi and I, conclude the negotiations, or, alternatively, that another person should participate on my behalf. After all, Kivi's partner had no idea as to how the royalty was obtained. To him, I was a broker entitled to a broker's commission.

The end result was that, instead of 1 percent or 1.5 percent royalty, I obtained just one quarter of 1 percent. What really hurt me in this case is not so much the money. Rather it is the fact that I was cheated of something that was really coming to me, and by somebody who was supposed to be my friend and whose financial security I created. Years later, I wrote a letter to Kivi, at least to put his ungrateful behavior on record.

The Khalda concession became the most productive area in the Western Desert of Egypt. Oil was first discovered there in 1984. Since then, more than twenty new oil and gas fields have been found on the Khalda Concession and new field discoveries continue to be announced frequently. There are even larger reserves of natural gas in the licensed area. A long pipeline has recently been laid and two hundred and fifty million cubic feet of gas started flowing in the last quarter of 1999. Six thousand cubic feet of gas are equal in calorific value to one barrel of oil. Therefore the throughput of two hundred and fifty million cubic feet of gas equals forty thousand barrels of oil per day. The production of gas will also increase considerably the production of oil. As this oil was associated with gas, its production was held back until the gas pipeline was ready. All in all, an oil bonanza. The biggest discovery in Khalda was announced recently in September 2003. It is believed to equal the combined total reserves of all twenty oil fields discovered there previously.

As I mentioned above, Bitton's Mleiha area also became very productive. First Denison Mines, a Canadian mining company, joined in. Then AGIP the Italian National Oil Company, as well as additional companies, became associated. They created the Mleiha consortium. Mleiha is probably the second-best producing area in the Western Desert, after Khalda.

Jack Bitton died several years after I first met him. I do not think that he lived to see the fruits of his efforts. The development of Mleiha came about after his demise.

But there is an unexpected, bizarre sequel to this story. Several years after Bitton's death, a series of programs was shown on Egyptian television. "Jack Bitton, the greatest Egyptian spy Egypt ever succeeded in planting in Israel." It seems that Jack Bitton was not Jewish, as he had presented himself to me, but a full-blooded Egyptian, either a Moslem or a Christian Copt. His travel agency in Tel Aviv was a cover for his spying activities. The Mleiha license was a reward for his services to the Egyptian state. A whole chapter in a book called *The Spies*, published in Israel in 2002, is devoted to Jack Bitton.

* * *

Although I was engaged in many projects, the 1980s were hard years for me, both personally and in business.

My wife had to undergo two cranial operations. They were performed in Israel and were not very successful. As a result, her right arm is weak, and, being a right-handed person, this causes her discomfort, inability to drive a car, and other problems.

I also had a serious financial disaster in the early 1980s. I invested substantial sums of money in Texas International Oil Company, headquartered in Oklahoma City. They reported immense discoveries of gas in Louisiana, which proved later to be a fraud. The company was recommended to me by my good and dear friend, Barth Walker, who was the most conservative of men, the best and the most astute oil lawyer in Oklahoma, and a man that I would trust with anything dear to me to this day. Barth Walker also must have lost a fortune in backing Texas International Oil.

The 1980s were also the decade in which my son Kobi founded Comverse Technology, which became the largest voice mail company in the world. In the final analysis, the best thing that happened to him was the fact that I could not help him at all financially, and he did it all on his own. Still, it was very difficult for me to watch him struggle.

As I have mentioned before, during the 1980s I was again twice involved in Ghana and in Nigeria, with the previously referred to Chief Tarka. However, the numerous changes of regime in Nigeria made each successive project unworkable, as it had to be dealt with anew with each incoming government.

I was also pursuing an interesting project of secondary recovery of oil in the Talara field in southern Peru. Talara was a very large oil field discovered in the nineteenth century, which produced during its lifetime upward of one billion barrels of oil. As this oil was produced by relatively primitive methods, very large reserves remained in the ground. Oil companies believed that a substantial part of that oil could be produced by secondary recovery methods.

Only a certain percentage of any oil reservoir (less than 50 percent) can be recovered by primary methods, because the rest of the oil remains "glued" to the sand grains. Secondary recovery methods, which usually include pumping great amounts of water under high pressure, can dislodge part of this "glued" oil and push it to the well head. Twenty to thirty years ago, it was believed that secondary recovery could be a tremendous source of additional producible reserves of oil in the world. But somehow or other this method does not work everywhere, and the ultimate quantities derived from secondary recovery are not very large.

Going back to Talara, MAPCO had a part of this tremendous field under license and was planning to install secondary recovery procedures there. As MAPCO could only get a quarter of this project for itself, Bill Thompson was gracious enough to give me their technical evaluation and studies. I had organized a group which included Foster Wheeler, a prominent American oil service company, and DEMINEX, a German consortium that included several German oil companies and the German government. On their behalf, I travelled back and forth to Peru during a two-year period to negotiate the contract with the Peruvian government. But the South American custom of mañana (tomorrow) did not allow the project to come to fruition.

Occidental Petroleum, a very large American oil company, which had a part of the Talara field under license for secondary recovery, installed enormous pumps, reported to be the largest in the world, to pump sea water into the Talara oil reservoir. This project may still be alive, but it is not a substantial money maker. Other companies, including MAPCO, have abandoned the secondary recovery project in Peru.

* * *

In late 1984, Bill Thompson left MAPCO and became an independent oilman. We had often discussed the possibility of joining forces when he became independent, and eventually we did. Bill, since his days in Burmah Oil, was in love with the prospects of oil explorations in Burma (which has since changed its name to Myanmar). One of the earliest discoveries of oil in the world, in the nineteenth century, was in Burma. A British company which made those discoveries was named Burmah Oil and was at one time one of the richest and most successful oil companies in the world. British Petroleum Oil Company was created early in the twentieth century by Burmah Oil, which was its largest stockholder, after the British Government.

No serious oil exploration had been done in Burma since 1939, when the country was invaded by Japan. After World War II, Burmah Oil returned to the country but was soon expelled as a result of nationalistic sentiments. A socialist, half-communist regime took over. It hated foreigners and it did not allow any foreign oil exploration to take place, until the beginning of the 1990s. As a result, Burma was probably one of very few countries in the world which had oil, but where for the last sixty years no serious exploration effort had been made.

The problem was how to penetrate the invisible wall that the Burmese regime had erected and to commence exploration in Burma. Bill and I were sitting in Houston one day in 1985 with a very wealthy industrialist friend of Bill Thompson's. He was ready to finance a serious exploration effort in Burma, if we would succeed in entering the country.

I had a brain wave and said: "Why don't we use the Romanians? They are communists like the Burmese and they also produce oil." To make a long story short, Bill Thompson and I travelled to Romania in October 1985 as the Guests of Rompetrol, the national oil company of Romania. The Romanians were in seventh heaven. They visualized green dollars, foreign travel, employment abroad, etc., etc. We said we did not need their money; the only thing we wanted from them was to open the door for us. A year-long Romanian effort to "open the door" brought no results. The Romanians did not even receive a reply from Burma, not even a polite "no." I have a whole file of one way telexes from Romania to Burma, with no Burmese reaction.

There was a humorous epilogue to the Romanian story. In the early 1990s, after the fall of the communist head of state Ceauçescu, Romania opened its doors to the west and invited western oil companies to come. I travelled to Romania with representatives of Clyde Petroleum, a large British independent oil company. During our meeting with the president of Rompetrol, Clyde Petroleum presented their credentials, and then I started introducing myself and A & T Exploration. The president of Rompetrol said: "Mr. Alexander, you don't have to introduce yourself and Mr. Thompson to us. We know you both very well."

* * *

In 1989 Bill Thompson and I had created A & T Exploration Company Ltd., the "A" standing for Alexander and the "T" for Thompson. The company exists to this day (2003). It is a small, private oil company, originally financed by the two of us and by some of our friends. The funds so raised were exhausted by the mid-1990s, and since then I have practically financed operations myself, spending more than one and a half million dollars from my own pocket financing the various ventures in Burma and in Russia, which I will describe briefly.

In 1990, Bill Thompson was invited by the Board of Directors to be the interim chairman of the largest bank in Oklahoma. There was a banking crisis in the USA and the former management of the bank had been removed. His original assignment was for three months, but eventually went on for over ten years. At the time of this writing in 2003, he is still connected to the bank. He advised the bank of his involvement in A & T and got their approval to continue. We still discuss various matters relating to our company at least twice a week, but all the practical activities of travel and management are carried out by me alone.

The only advantage deriving to me from Thompson's association with the bank is my ability to state that "my partner is the chairman of the largest bank in Oklahoma."

The main purpose of creating A & T was exploration in Burma. We were part owners of two exploration licenses there. One other partner is a British company and Dr. Zevi Dinstein, and the group which he heads. So the circle has been closed. Zevi Dinstein, who had been my boss for many years and whom I have mentioned so many times in this book, was a partner in our exploration effort there. The American sanctions against Burma, due to human rights abuses there, came into effect a few months after we got our original license, and they have made life very difficult for us. More than 95 percent of the oil companies in the world are in America, and none of them would touch Burma with a ten foot pole because of the US sanctions. We have most probably a large undeveloped oil field in our licensed area. It was originally discovered by Burmah Oil in 1962, just before they were kicked out of the country. The well discovered a producing horizon three hundred feet thick, which may contain millions of barrels of oil or gas. The well was never tested, as Burmah Oil was expelled before the testing could commence.

Due to these American sanctions, it has been almost impossible for us to find partners to drill, test and develop this oil field. Therefore, I have regretfully decided to stop our investment in the Burmese projects as of January 1, 2003.

* * *

The other projects that A & T was involved in were in Russia. Let me state from the beginning the conclusion I came to quite early on. As far as Russia is concerned, the oil companies in the world should be divided into two groups. The first group includes the major oil companies of the world who cannot afford not to be in Russia, due to the very large undeveloped oil and gas potentials and reserves in the country. The second group, including all the other oil companies in the world, cannot afford to be in Russia because of the complicated and difficult terrain and climatic conditions, the cumbersome bureaucracy and the prevailing dishonesty. A company that is able to look ten to twenty years ahead, that can make large investments and forgo profits for many years, should be in Russia. All the others should stay out. I include in the term "Russia" also the now independent states which comprised the former Soviet Union.

Now to specifics of our activities in Russia. We were involved in a large, three hundred million barrel oil field in Tomsk, Siberia. I actually have a signed agreement to develop this oil field, dated in early 1990. Over ten years have passed, and I am convinced that the situation in this oil field has not changed. It is still waiting for development.

During my numerous trips to Tomsk, I came to the conclusion that there is another natural resource there, which is readily available on the surface, without any need for drilling. This resource is the hundreds or even thousands of scientists who lived in Tomsk and were suddenly grossly underemployed. They still worked in their scientific institutions, but their budgets had been cut or had disappeared completely. The whole purpose of their professional life, of developing materials, techniques, and inventions for the Russian military effort, had disintegrated.

Tomsk was a closed city during the Soviet regime. You needed a special permit to enter and leave the city, as it was totally dedicated to the Russian war effort.

I reminded myself that the Russians had actually launched the first satellite (Sputnik) before the Americans did. Consequently the country must have a very large reservoir of brain power. In order to check the validity of my assumptions I formed a scientific delegation, composed of Professor Peter Morse, the head of the largest university computer department in England, and two other scientists, to go to Tomsk and investigate the possibilities there. They made repeated visits during the following few years and wrote very enthusiastic reports. Part of one report is quoted below.

> Why Tomsk:
>
> 120 man days have been spent by the independent consultant team in Tomsk. All universities and a selection of research centres and government institutions have been visited...Meetings have been held with the most senior staff. The database generated on the Tomsk activities for A & T is one of the most comprehensive in existence. Based on the international experience of the consultants, who have worked in Russia, the US, throughout the Middle East, Africa, China, India and Europe, it is considered that the Tomsk complex has one of the highest capabilities and industrial potential ratings in the world in cost/performance and quality terms... Of outstanding importance were the scientists and engineers themselves... The population of Tomsk is entirely... European and the standard of English is high. This is a unique collection of some of the most creative and intelligent technologies that I have met.
>
> The Tomsk industrial and scientific city (formerly one of the Soviet Union's "closed cities") with its population of technologists is only paralleled by Silicon Valley, Houston, Massachusetts and the Moscow region. It does, however, have one fundamental difference. It lies in one of the richest areas in the world with limitless energy resources, a massive petrochemical industry and the world's largest timber resources, as well as substantial minerals, gold, uranium and diamonds. The skill of the Tomsk engineers will be turned to exploit this wealth and their future will be secured by it. This has probably the greatest potential for your investment and plans.
>
> *Professor Peter L.R. Morse, head of the School of Computer Science & Information Systems Engineering, University of Westminster, the largest computer science faculty in Europe, in a report to A & T dated 4 January 1994.*

There is a large investment group in London called the Framlington Group of Funds. In 1994 they created a fund for investment in Russia which they called The Framlington Russia Fund. The president of the

Russia Fund, Gary Fitzgerald, travelled with me to Tomsk and was enthused with what he saw. He recommended to his Board to invest two million dollars in our Tomsk scientific initiative, for a 20 percent interest, giving a valuation of ten million dollars to our project. His board did not accept his recommendation. They decided that our project was too ambitious and that, initially, they should invest only in unsophisticated bread and butter ideas. Several months later I succeeded in arousing the interest of the British Aerospace Company in the Tomsk project. The head of research and development of BAC travelled to Tomsk and commissioned several research projects through us. Following the British Aerospace involvement, I succeeded in interesting Israeli Aircraft Industries (IAI) in the Tomsk idea. Israeli delegations from IAI and Rafael (Israel's Weapon Development Authority) visited Tomsk, and delegations from Tomsk visited Israel. With the economic collapse of Russia in 1996, this whole effort became dormant. Although I spent more than half a million dollars of my personal funds on the Tomsk scientific idea without any returns, I am glad and proud that I made this effort. I liked and respected the people of Tomsk and wanted to help them. They were caught in a "hurricane" which was not under their control. Tomsk science was a great idea.

The last project in which we were involved was diamond exploration in Krasnoyarsk, Siberia. It was an interesting project with great potential. We even had De Beers interested in it, and their Moscow representative flew in to see us in London. We had British partners in the diamond venture, and together we formed a Canadian public company called Siberian Pacific Resources that was listed on the Vancouver stock exchange. The economic collapse in Russia in 1996 put an end to this effort as well.

* * *

Before closing, I would like to tell the story of Chief Lawson, with whom I first became acquainted in 1992. He was the only Nigerian I have ever met who was a perfect gentleman and a most impressive individual. He was so different from all the Nigerian notables with whom I had dealt in the course of thirty years that it seemed as if he came from another country and from a different continent.

He was a barrister, educated in England, the owner of the largest bottling plant in Lagos, chairman of a large bank, and involved in many other enterprises. He was very interested in the oil business and familiar with it. He had even participated in financing the drilling of some oil wells in Gabon.

After many discussions I had with Chief Lawson regarding various facets of the oil business, I suggested to him the formation of a new and completely different type of oil company in Nigeria which would have wide Nigerian public participation. No such "animal" existed there at the time, nor does it even today. I thought that the people of Nigeria should have an opportunity to benefit directly from the oil wealth of the country. Also, the Nigerians liked to gamble, like people elsewhere, and I believed that such a company would be met with great enthusiasm in Nigeria. I suggested that Chief Lawson should first invite several of the most prominent and successful individuals, representing the various tribes and regions in Nigeria. Only after that should the public be invited to join. The company would then be listed on the Nigerian stock exchange. I, on my part, undertook to bring in a western oil company that would be both an investor and a provider of the necessary technical expertise. Chief Lawson loved this idea and, within a month, a company was formed and an agreement to join was reached with Premier Consolidated Oil Fields, a large independent oil company in London.

A contract among all the parties involved, along the lines described above, was signed on February 9, 1993, and Nigerian Premier Oil Ltd (NPOL) came into being. Premier experts chose two exploration blocks, one onshore and one offshore Nigeria, which they considered to be most promising.

Nigerian Premier Oil was so well accepted in Nigeria that the government of Edo State, where the chosen blocks were located, decided to participate in the new company with an interest of 20 percent.

In 1995, Chief Lawson underwent an operation in Washington, D.C. While recuperating at his daughter's home in Washington, he suddenly collapsed and died. The venture practically died with him. His son, Kolapo Lawson, who is as straight, intelligent and able as his father had been, still tries to resurrect the project, but without much success so far.

* * *

To describe the various and strange "ventures" which were presented to me in the last twenty years would probably fill several books. Many of them I no longer recall and, frequently, when I come across some letter in one of my files, it takes me a long time to remember what it was all about.

Our place in London, to which Rachel and I returned in 1977, became a meeting place for all kinds of individuals, belonging to diverse tribes, nations and countries. There were princes, ministers, chiefs of staff and high ranking officials among them. They came primarily from Africa and some from South East Asia. Many of them had the misfortune to be jailed soon afterwards, following the numerous coups in their countries. A few were executed. In addition to the Africans and Asians, I was also often approached by strange western entrepreneurs with weird ideas which did not have any chance of success. But I was always prepared to give people a sympathetic preliminary hearing.

I look back on a long period, very difficult at times, very risky from a business point of view, and sometimes even fraught with personal danger. The last twenty-five years have been totally different from the years that I spent at the Israel National Oil Company. There I had an office, secretaries, drivers and all other amenities of the head of a company. Furthermore, while serving at I.N.O.C., I felt that, in spite of all the constraints and difficulties, I still had the State of Israel standing behind me. During the years of independent activities, after I left I.N.O.C., everything depended on me and me alone. I had to fight a constant war of survival single-handedly. It was a completely different world.

On many occasions, several of my friends, hearing about the many strange events described here, urged me to put them down on paper. Finally I accepted their advice and wrote this book. While writing the various chapters, the excitement and joy of entering this fascinating field of oil exploration came back to me, together with the memories of all the disappointments, missed opportunities, and frustrations. Still, the writing of the book gave me the opportunity to review and summarize my activities in Israel's oil world, a period of which I feel very proud.

Looking back I think I have led a very interesting life. It was a life full of difficulties, many failures and few successes which had to pay for all the other failures and thank God they did. My main problem was always that I was a one-man army, fighting in this difficult and complicated world. Still I would not change it for anything.

EPILOGUE

I returned to Israel in 1958 after four years' service in the Israeli Ministry of Defense Purchasing Mission in New York. I was greatly influenced by the economic scene in America and by my studies at Columbia University. While searching tirelessly for oil in Israel proper, and later, outside its borders, and viewing the scene in Israel, I came to the conclusion that the economic emphasis within Israel was completely wrong. To name but a few examples: Israel's industry was based then on textile plants, which thrive on cheap labor, and steel production in a country that has no iron ore, no coal and very costly electricity. In the agricultural field, Israel introduced cotton-growing, which requires large quantities of water in a country where water is scarce and very costly.

I believed, and stated, already then, that Israel should devote its efforts to the field of electronics (the high tech of the 1950s, 60s and 70s). My reasoning was that this was the industry of the future. Its "raw materials" were brainpower, which Israel has in abundance. It does not require natural resources, which we do not possess. It can be developed in a garage or an attic, and it thrives on entrepreneurial spirit, in which Israel excels. And finally, it is not subject to political pressures.

I am happy that this transformation eventually took place in Israel, although it took thirty to forty years to develop. I am especially proud that our two children participated in a big way in this transformation. As described in chapter 9, our son Kobi founded, and is heading, one of the largest and most successful high tech companies in Israel—Comverse Technology. Comverse is the world's leading provider of software and systems enabling, network-based, multimedia enhanced communication services, and is a world leader in voice mail, employing thousands of people in Israel and abroad.

Our daughter Shaula, an IBM Research alumni, founded SMARTS, a successful software company solving complex IT management problems through innovative technology. SMARTS customers include the world's largest enterprises and service providers. Thus the move from traditional oil exploration in Israel to the industries of the future has been accomplished in our family in one generation.

This is the "OIL" of the 21st Century.